HOW ANIMALS MOVE

JAMES GRAY

How Animals Move

BY

JAMES GRAY

*Fellow of King's College,
and Professor of Zoology in the
University of Cambridge*

ILLUSTRATED BY

EDWARD BAWDEN

CAMBRIDGE

At the University Press

1959

CAMBRIDGE UNIVERSITY PRESS
Cambridge, New York, Melbourne, Madrid, Cape Town,
Singapore, São Paulo, Delhi, Mexico City

Cambridge University Press
The Edinburgh Building, Cambridge CB2 8RU, UK

Published in the United States of America by Cambridge University Press, New York

www.cambridge.org
Information on this title: www.cambridge.org/9781107621374

First edition 1953
First published 1953
First issued in this format 1959
Re-issued 2013

A catalogue record for this publication is available from the British Library

ISBN 978-1-107-62137-4 Paperback

To

Anthony James and
Sarah Ann

Contents

List of Plates

List of Figures

9

Preface

THIS book represents the substance of six Christmas lectures delivered to a juvenile audience at the Royal Institution. The lectures were designed to illustrate the more striking features of animal locomotion without assuming previous biological knowledge. Wherever possible, living animals were allowed to tell their own story, and there can be little doubt that they succeeded where the spoken word or blackboard diagram would have failed.

Five of the lectures were concerned with relatively well-established facts. The sixth lecture was of a different type, for it enabled the audience to see research workers in action and hear from them, at first hand, something of the problems which biologists are trying to solve and how they set about their task. Dr F. S. J. Hollick and Dr J. W. S. Pringle demonstrated the methods they have developed for investigating the flight of insects; Mr D. O. Sproule provided an 'artificial' bat avoiding obstacles by supersonic echo-sounding, and Dr H. W. Lissmann demonstrated the electric fields which he has recently shown to exist round the bodies of certain fish. These demonstrations are referred to at various places in this book but not, I fear, in such a manner as reflects the striking impression made by them in their original form.

It is impossible to express adequate thanks to all those who helped in the production of these lectures. Particular gratitude is, however, due to the British Museum (Natural History) for the loan of many valuable specimens; to the Zoological Society for providing a unique series of living animals; to London Films International for the loan of the film *Around the Reef* and to Gaumont British Films for a number of slow motion films of human athletes; to Messrs Lines and Messrs Hamley, and the International Model Aircraft Company for the loan of mechanical models; and to Messrs Cinema Television Ltd and Messrs Dawe for electrical equipment and Messrs Gerrard for the loan of various specimens.

On the staffs of the Royal Institution and of the Zoological Department, Cambridge, fell the major task of preparing and assembling the demonstrations. It is impossible to thank each person individually, but three names must be mentioned: Mr L. Walden of the Royal Institution and Mr K. Williamson at Cambridge who for many weeks spared neither time nor effort, and Dr H. W. Lissmann, who not only designed and set up many of the exhibits, but who persuaded so many animals to do just what was expected of them at just the right time, and later on read the proofs of this book. An author's acknowledgements to his publishers may be somewhat formal, but in this instance I would like to express particularly sincere thanks to Mr F. Kendon of the Cambridge University Press not only for a revision of the original manuscript but for all that he and his colleagues have done during the publication of this book.

J. G.

1

The Machinery of Animal Movement

ONE thing which obviously distinguishes animals in general from other forms of life is a power they have of moving themselves from place to place. Different animals move in different ways – some creep along like a worm, some walk by means of legs, others paddle themselves through the water, or fly through the air. All these movements appear to be very different from each other and different too from the movements of such things as cars, ships, and aeroplanes. In this book I am going to try to tell you how far such impressions are true, and to help you to watch moving animals with increased understanding and pleasure.

If an animal is to move about in an orderly way it must, like a car, be provided with an engine, an efficient steering gear, and brakes; but before we look for these parts in an animal's body we may as well first of all consider just what is required from the engine of a motor car if it is to drive the car forwards. We all know that a car, with its engine switched off, will not start to move over a level surface by itself: to make it move we have to push it with our hands (Fig. 1). We can only push it with our hands if, at the same time, we give an equal and opposite push against the ground with our feet. So far as the car is concerned, if it is to move forward it must be pushed. If it gets no such push the car is an inert object; that is, one which must be forced to overcome its inherent incapability to move by itself. In the language of science we say that a car has *inertia*, and that this inertia is overcome by applying an external force. A force gives the car the energy necessary for movement; it will then go on moving by itself until this energy is taken away from it by another

FIG. 1. *We can only push a car forward if, at the same time, we give an equal and opposite push against the ground with our feet*

external force acting in a direction opposite to that which set the car in motion; such a force is usually applied to a car in the form of friction between the tyres and the ground. When we get into a stationary car, switch on and start up the engine, and put in the gear, the car moves forwards, but this forward movement still depends on exactly the same essential principle as operates when we push it with our hands; the only difference is that the back wheels and axle replace our feet and hands: the tyres of the back wheels press backwards against the ground and by this means an equal but forward force is applied by the axle to the body of the car. If instead of pressing against the ground the back wheels skid freely, no forward pressure develops against the chassis, and consequently no forward motion is imparted to the car. We can illustrate these facts by placing the hind wheels of a model electrically propelled car on a horizontal surface, so arranged that the surface moves backwards against a spring whenever a backward force is applied to it; the front wheels of the car rest on solid ground (see Fig. 2).

FIG. 2. *The tyres of the back wheels press backwards against the ground and by this means an equal and opposite force is applied by the axle to the body of the car*

As soon as the engine is switched on, the car begins to travel forwards, but, at the same time, the platform moves backwards. The force driving the car forwards is exactly equal to that driving the platform backwards against its restraining spring. Later on we shall do just the same experiment with different kinds of animals and we shall get the same fundamental result: an animal can only propel itself forward by pushing backwards against its surroundings (Figs. 3, 4). An animal on the land must push backwards against the ground; an animal in water must push backwards

FIG. 3. *A man propels his body forwards by pushing backwards against the ground with his foot. When he steps off a platform mounted on springs, the propulsive force can be measured by observing the extent to which the spring* S *is compressed*

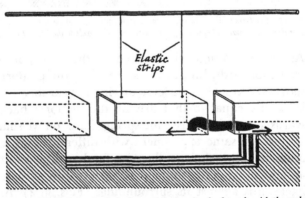

FIG. 4. *A leech pushes its body forward by pressing backwards with the sucker at the hind end of the body. The propulsive force applied to the body can be measured by observing the bending of the elastic strips*

16

against the water and a flying animal must push backwards against the air. In every case, the force available for moving the animal forwards is exactly equal to that with which the animal pushes backwards against its surroundings.

In cars, ships, and aeroplanes, the backward pressure against the surroundings is exerted by moving parts of various kinds but all called, as a class, 'propellers' – the back wheels of a car, the 'screws' of a steamer or an aeroplane. A 'propeller' of some kind or other exerting a backward force against the surroundings is an essential part of all self-propelling systems, and consequently (as an animal is always a self-mover), we can start our study of animal movements by looking for this all-important part, remembering that the 'propeller' must act against the animal's surroundings – the outside world – and must move relatively to the rest of the animal. Often the 'propellers' can quite easily be identified: the legs of a dog move to and fro on the body, and during their backward movement press against the ground; the wings of a bird move up and down, beat against the air. But sometimes the propellers are less obvious. A snail appears to creep along without changing its shape, but if we look carefully at its under surface, by allowing the snail to creep up or along a sheet of glass, we find that some parts are attached to the glass, whilst other parts are moving. Every animal moving itself must, in fact, obey these two rules – it must be constantly changing its shape, and it must press backwards against the outside world.

When we compare the propellers of animals with those of self-propelling machines made by man, we find one very striking difference. Except the jet propellers of modern aircraft, all the inventions of mankind depend upon rotation; the moving parts operating against the outside world are constantly turning in circles, relative to some other part of the machine. If one part is to rotate about another part in this way, the two parts must be separate from each other. The moving part turns upon an axle or bearing, across which no permanent connexions between the two parts can pass; otherwise they would soon get twisted and broken. In an animal no such arrangement is possible. Every part of the live body is connected to the rest by blood vessels and nerves, and these would quickly be destroyed by continuous twisting. Nature never uses a true wheel; she uses rods or levers, and these can move up and down, or from side to side, but can never make complete revolutions about a stationary axis. Never to be able to make use of a wheel-form seems at first sight a serious disadvantage; but we must remember that a wheel on its axle is really only a series of levers, coming into action one after another. As you see in Fig. 5, a six-spoke wheel rolling along on its rim can be regarded as six legs each ending in a foot; when we turn the axle in the direction of the arrows only one foot is in use at a time, but as soon as the toe of one foot leaves the ground, the heel of the next foot comes into touch with the ground; having pressed against the ground, each spoke turns round on the axis and in due time is ready to come into

use again at the right moment. As a propeller, however, it does not matter what a spoke (or leg) is doing
after it leaves the ground so long as it is ready to take
up its propeller duties again in the right place and at
the right time. For example, we can replace six legs
arranged in the form of a wheel by two legs each

FIG. 5. *A six-spoke wheel rolling along on its rim can be regarded as six legs
each ending in a foot. Only one 'foot' of the wheel is in use at a time, but as soon
as the toe of one foot leaves the ground, the heel of the next foot comes into contact
with the ground. We can replace the rolling wheel with its six spokes by two
legs; if each swings forward when it has performed its task as a propeller, it
comes into action again at the right moment*

swinging forward again as soon as it has done its duty
as a propeller – a motion obviously very similar to that
of our own legs. The limb is turning about its upper
end, just as each spoke of a wheel turns about the axle –
the difference between our legs and the spokes of the
wheel is that our leg turns *forwards* about the hip joint
when the foot is off the ground, and *backwards* when
the foot is on the ground; but the spoke of a wheel
moves continuously in one direction in a circle. Mech-
anically the propulsive effects are the same.

Instead of using only two levers swinging alternatively backwards and forwards, we can arrange a
series of legs one behind the other and coming into

action one after another. This is, in fact, one of Nature's commonest arrangements, especially for animals with large numbers of legs. Each foot touches the ground a moment before the one just ahead of it, the legs move rather as the keys of a piano move when a finger is drawn along them. Fig. 6 shows the pattern of leg movement in a large millipede from South Africa. This animal has about four hundred legs, and these give the impression of 'playing scales' on the ground as the animal walks along. We are asked from

FIG. 6. *A large South African millipede has about four hundred legs. Each foot touches the ground a moment before the one just ahead of it, and the legs move as the keys of a piano move when a finger is drawn along them*

time to time, 'Which leg does a centipede (or millipede!) move first?' The answer is that all legs start together just as the spokes of a wheel do: but the legs of the centipede are kept in the right position in relation to each other by a very complex piece of nervous 'machinery', whereas the spokes of a wheel are fixed together by a mechanical axle. We can sum up the whole situation by saying that engineers use levers fixed together to form rigid revolving wheels; Nature gets the same mechanical results by moving her levers up and down or from side to side. Or, if we like, we can say that a man propels himself by two spokes, a dog by four, an insect by six, a spider by eight, and a centipede by a very large number!

In this book we shall be dealing almost always with the movements of fairly large, common animals, and we shall find it fairly easy to regard the bones as levers and the muscles as active engines. It is worth remembering, however, that such animals as this are the *latest* of Nature's designs. In the early stages of her experiments Nature worked on a much smaller scale; long periods of time passed before the principle of rigid levers, driven by muscular engines, was adopted. The principle of muscular levers was preceded by two other much less successful designs.

If you look carefully at the surface of the mud at the bottom of a shallow muddy pond, you may sometimes see very small white spots, which, examined under the microscope, turn out to be little irregularly-shaped creatures called *Amoebae*. These animals have no proper limbs, but move along by squeezing out blunt processes from the surface of the body. One of these processes attaches itself to the surface of the mud and the rest of the body flows either towards or away from it; then a new process of 'pseudopodium' forms, and the cycle is repeated. An amoeba is a very small animal – about one-twentieth of the size of a pin's head – and it moves along at a rate of not usually more than half an inch an hour, or one foot in a day. Movement of this kind – amoeboid movement – has not been developed as a means of propulsion for larger animals. But Nature has not forgotten it altogether; it plays a very important part in the life of such animals; by amoeboid movement the white cells of our blood can flow around and swallow up foreign particles which may

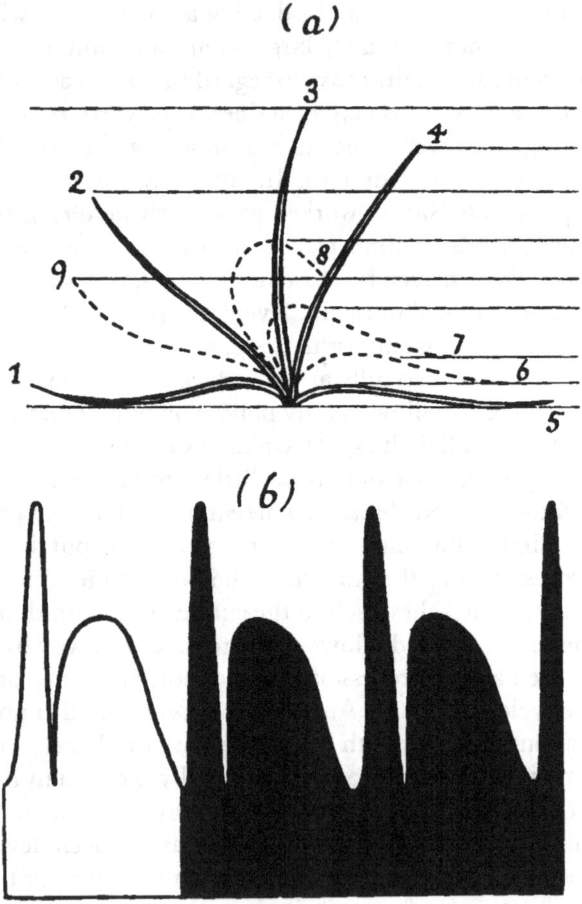

FIG. 7. *Cilia sweep backwards like the pliant lash of a whip, but forwards like a stiff rod. Thousands of cilia work in relays creating an impression of waves passing over the surface of the body, like waves passing over a field of corn in a gust of wind. In Fig. 7a a single cilium is beating in the plane of the paper; the*

have found their way into our bodies.

After very slow movements of the amoeboid type, we come to a rather faster way of getting about, characteristic of many of the microscopic creatures found in almost any drop of pond water or mud. When we magnify such animals we find that the surface of their bodies is covered by a felt of very fine hairs, known as *cilia*. These hairs are in constant motion backward and forward; they sweep backwards like the pliant lash of a whip, but forwards like a stiff rod. The motion is not unlike that of an arm when bowling a cricket ball, except that it is repeated ten or twenty times a second! Thousands of these hairs cover the body of the animal and they usually work in relays, creating an impression of waves passing over the surface of the body, like waves passing over a field of corn in a gust of wind (Fig. 7). When we watch these cilia under the microscope they seem to us to move at great speed, but when we remember that each cilium is in fact only about $\frac{1}{1000}$th of an inch long, we realize that their speed of movement is really very small and that the speed with which the animal moves is also very small – perhaps 15 ft per hr or 120 yd per day. The forces exerted by a cilium are minute, and it is therefore not surprising to find that none of the larger animals use cilia to propel themselves. Ciliary movement (like amoeboid movement) has been used for another

forward stroke is seen in positions 1–5, the backward stroke in positions 6–9. In Fig. 7b a very large number of cilia are beating in a plane at right angles to the paper; the sharp spikes are cilia performing their forward propulsive stroke and the lower waves are cilia in their backward strokes. In life, the waves move along just like the waves of the legs of the millipede in Fig. 6

job. In the higher animals, cilia are used for creating currents of water or other fluids over surfaces which have to be kept clean (for, as we shall see later, a propeller cannot drive a submarine or an animal forwards through water without at the same time driving water backwards in its wake). The walls of the tubes that convey air through our nose to our lungs are covered with cilia, and the mucus that collects in our nose or throat when we have a cold is brought there by the joint activity of millions of cilia, all driving fluid away from the lungs. But ciliary movement as a method of locomotion is only of use to very small creatures.

The movement of all the larger animals depends on muscle fibres, for these structures are by far the finest and most powerful of all Nature's engines. The muscles of our bodies are made up of a number of these fibres bound together, and each fibre, though sometimes several inches long, is not more than a very small fraction of an inch in diameter. So long as its fibres are at rest, the muscle is soft, flabby, and can easily be stretched. Each fibre is supplied by a nerve coming from the brain, and when impulses pass down this nerve the whole character of the muscle changes. It shortens in length and strongly resists any attempt to prevent this happening. It changes, in effect, from a piece of perished rubber to a stretched steel spring. As soon as the stimulus from the nerves ceases, the muscle becomes limp again; but it does not stretch itself out to its original length; to regain length the muscle must either be pulled out by the contraction of another muscle or extended by some other force not its own.

In some animals (a leech or earthworm for example), we can see the shortening of the muscles as the animal moves along, because the shortening of the

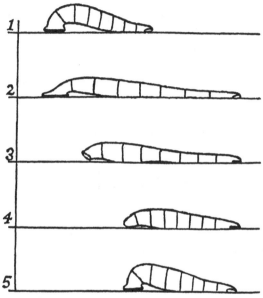

FIG. 8. *A leech moves by using two sets of muscles, longitudinal and circular. At each end of the body is a sucker. When the hind sucker is fixed to the ground, the circular muscles contract and the body extends forward (1, 2); the front sucker is then fixed to the ground (3), the hind sucker released and the body drawn forward by the contraction of the longitudinal muscles (4); the hind sucker is then fixed to the ground again (5)*

muscle fibres causes a shortening of the whole body. These changes are seen very clearly in a leech if a series of lines has been painted across its back with chinese ink. The animal moves by means of two sets of muscles – one series (longitudinal) running along

25

the body and the other (circular) running round the body. At each end of the body is a sucker, which the animal can attach firmly to the ground. When the front sucker is attached, the hind sucker is free; when the front sucker is fixed to the ground the longitudinal muscles are fully stretched and the circular muscles fully shortened. But now the longitudinal muscles shorten and the circular muscles elongate, and as this happens we can watch the marks on the back of the animal move closer together as the animal is drawn forwards towards the front sucker; the whole body becomes shorter but thicker. As soon as the longitudinal muscles have fully shortened, the hind sucker attaches itself to the ground and the front sucker detaches itself. Simultaneously, the circular muscles shorten and the longitudinal muscles are stretched and the whole body is pushed forwards; it becomes longer and thinner, and the marks on the back move further apart. After this, the front sucker again attaches itself to the ground, the hind sucker detaches itself, and the whole succession of events is repeated (Fig. 8). Essentially the same method of moving forwards is to be seen in the earthworm. But the body of the worm is much longer than that of a leech, and the longitudinal muscles do not shorten all along the body at the same time: the longitudinal muscles over one region shorten whilst the circular muscles are shortening at another; and the earthworm's grip of the ground is not maintained by suckers but by fine bristles which act somewhat like grappling-irons. How do the leech and the earthworm exert backward forces against the ground? Let us allow the animal to move over a suitable version of the 'Lissmann' bridge, shown in Fig. 4. We shall see that, as the animal moves forwards, forces of from two to eight grams are exerted against the ground. No worm could exert force against perfectly smooth ground, and no forward movement could then take place.

In an earthworm or leech the longitudinal muscles shorten on the two sides of the body at the same moment, and the changes in shape of the body resemble those of a rubber tube alternately stretched and released. But in other animals, the muscles on one side grow shorter as those on the opposite side grow longer. In such animals the part of the body concerned does not shorten and lengthen but bends alternately to one side or the other. The muscles all down one side do not shorten simultaneously, but in a definite order, and the result is that the whole body is thrown into a series of wavy curves. These curves are very obvious when the marine worm *Nereis* is swimming in water (Fig. 9), and we shall see them again in eels and snakes.

We should note, in passing, that the rhythmical movements of a leech, earthworm, or *Nereis* are brought about by two groups of muscles, each exerting an effect opposite to that of the other. When the circular muscles of a leech shorten, they automatically stretch the longitudinal muscles, and vice versa: similarly, when the longitudinal muscles contract on one side of a *Nereis* they stretch the muscles on the other side. Two sets of muscles acting in this way form what is called an 'antagonistic pair'.

27

As you know, earthworms and leeches are soft bodied animals. Most more active animals, however, have either a hard jointed armour on the outside of their bodies (e.g. insects, crabs) or a rigid jointed skeleton within it (e.g. fish, dog, horse). In all these

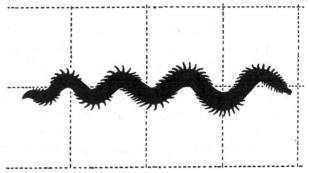

FIG. 9. *The marine worm* Nereis *swims by throwing its body into waves which travel from tail to head. As its body is covered with many small paddles, the animal moves in the direction in which the waves are travelling, i.e. head first*

more active forms, the ends of each muscle are attached to these rigid structures and the muscle turns the part of the body concerned about a rigid hinge. We see this turning process very clearly in our own legs and arms, for each muscle is attached to two bones hinged together by a joint; when a muscle shortens across the hinge the latter opens or closes according to the side on which the muscle is situated (Fig. 10). When the *biceps* of our arm shortens, the wrist is brought nearer to the shoulder and the arm is flexed; when the *triceps* on the hinder side of the arm shortens the arm is extended. Notice that when the biceps

28

shortens, the triceps is automatically stretched, and vice versa; by using these muscles alternately the arm can be flexed and extended time after time. The biceps produces a movement opposite to that of the triceps, and the two muscles thus work against each other; they are in effect *antagonistic* to each other. We

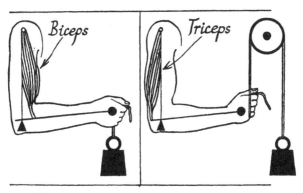

FIG. 10. *When a muscle shortens across a bony joint, the latter opens or closes according to the side on which the muscle is situated. When the* biceps *of our arm shortens the arm is flexed and, if the tension of the muscle is sufficient, the hand can exert an upward pull against a weight; when the* triceps *shortens the arm is extended and the hand can exert a downward pull. The shortening of the biceps automatically lengthens the triceps and vice versa; both muscles operate the forearm as a lever whose fulcrum is the elbow joint*

shall find that pairs of antagonists are responsible for the propulsion of almost all the larger animals.

We have now reached a stage at which we can begin to regard each member of an antagonistic pair of muscles as an engine, whose task it is to turn an animal's legs backwards or forwards as the engine of a car turns the car's back wheels. The muscles operate the bones as levers and because this action is by far the most important principle in animal movement, we shall have to consider it with some care. First, we must remember that a muscle cannot pull on one bone without pulling in the opposite direction against the bone which forms the other side of the common hinge. Further, the muscle cannot exert a tension between its two ends without pressing the two faces of the hinge together with a force exactly equal and parallel to its own tension. When a muscle exerts its effort, two forces act on each bone, the pull of the muscle itself and the pressure acting on the surface of the hinge; such pairs of forces are known as turning couples. The turning couple applied to one bone is always exactly equal but opposite to that acting on the other bone. As you know, any rod when subjected to a turning movement can act as a lever.

Levers, as used in everyday life, are usually devices whereby one end of a rod is made to exert a powerful force when relatively small turning forces are applied to the rod. To be used in this way, the rod must rest on a fixed pivot (the fulcrum) situated nearer to the end at which we want to develop the powerful force than to the end at which we apply our own efforts. If the pivot or *fulcrum* is ten times nearer to the 'load' than to the 'power', the force developed against the load is ten times that applied to the other end of the lever. This is known as a 'Lever of the First Order' and is useful when we want to move a heavy weight and do not much mind if the rate at which we move it is slow, as when we use a crowbar for lifting a heavy weight; if the lifting force applied to the weight is ten times that exerted on the lever by our hand, the speed with which the weight rises is ten times less than that at which our hand moves downwards. A few of our muscles operate bones as levers of this first order, but most of them differ from a crowbar in one very important respect (Fig. 11). If we call the region between

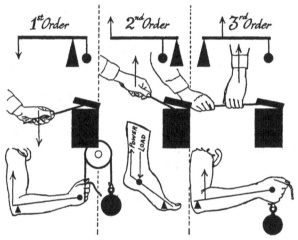

FIG. 11. *Our muscles operate our bones as levers. In nearly all cases the power arm of these levers is shorter than the load arm*

the load and the fulcrum the *load arm* of the lever, and the region between the fulcrum and the muscle, the *power arm*, we find that the power arm is nearly always very much shorter than the load arm: this is the reverse of the arrangement when a crowbar lifts a heavy weight. The advantage for our muscles of a short power arm is that muscles develop their greatest power when they shorten very slowly whilst

31

developing very high tension. The body uses these powerful, but slow-moving forces, for exerting smaller but much more rapidly moving forces at the end of the load arms of the bones.

In mechanics it is often convenient to divide levers into three orders: first, second, and third, according to the relative positions of the load, fulcrum, and power; but for biological purposes it is sufficient to remember that all bones act as levers when a muscle shortens across their common hinge, and that, almost always, the power arm is much shorter than the load arm. The tension in the muscles is therefore usually many times greater than the forces that our bones exert against the outside world. An exception to this rule occurs when we raise the whole weight of our body on to our toes by means of our calf muscles: here the toes form the fulcrum, the weight of the body acts on the ankle joint, and the powerful calf muscles pull on the heel. This is clearly a lever of the second order. But if the ankle joint is fixed in space, leaving the toes and heel free to move, the muscles of the calf can then operate the foot as a lever of the first order.

It is very important to remember that the practical result produced by the shortening of any particular group of muscles depends on the *external* forces acting on the body. If we push our arms out against a rigid wall, our body moves away from the wall, but if we extend our arm against something that is free to move, our body stays still and the object is moved away from us. We shall see later how essential it is that the foot of an animal should be fixed to the ground if the animal's body is to move along.

Each of our limbs comprises quite a number of bones and there are many pairs of antagonistic muscles. Clearly, the whole mechanical system of levers is extremely complex. However, we can simplify this complicated picture in several ways. In the first place, we shall be mainly concerned with the forces which animals exert against the outside world: we shall be less interested to know how the energy of the various muscles is applied to the individual levers inside the body. In other words, we shall be more interested in the parts that are comparable with the driving wheels of the motor car or the screw of a steamship than in the gearing that connects these parts to the engines.

Having seen that the tensile engines (or muscles) drive the propellers of so many animals by making the bones work as levers, we can look for a moment at the 'throttles', which control the amount of energy coming from the engines. So far as we know at present, each individual muscle fibre either remains inactive or exerts its full amount of power when it shortens. If an animal moves a joint slowly and against very little resistance, only a few fibres of the muscles concerned are brought into action by the nerves; when more strength is required, more fibres are brought into play. If we like to regard each fibre as comparable with one cylinder of a car's engine we increase our muscular efforts by using a larger number of cylinders instead of by putting an extra load on those that are already in action. As no muscle fibre can come into

C 33

action at all without receiving electrical impulses from a nerve from the brain, we can say that the 'throttle' of our muscular engine is controlled by the brain, which is really the 'driver' of our engines.

We can, however, carry this comparison a little further. As you know, the driver of a modern car or aeroplane does not rely on his ears to know whether the engine is running satisfactorily, or on his nose to know whether the oil is circulating properly. He has a revolution-counter, an oil gauge, and a speedometer, and the aviator has instruments for telling him how high he is, and how fast he is sinking or climbing. A careful driver or pilot relies a good deal on these instruments for exercising proper control over the movements of his machine. In the same way, the brain of an animal is informed of the state of tension of each of the muscles, and the relative position of one bone to another, by certain sense-organs: the 'instruments' which give the brain this information lie in the muscles or joints themselves; they are very small, but each one is connected to the brain by a nerve. When a muscle develops tension or changes its length, these little instruments (or proprioceptor sense-organs) generate electrical impulses in the *sensory nerves*; these impulses travel to the brain, and are taken into account when the brain adjusts the amount of power to be used, by sending impulses down the *motor nerves* to change the number of muscle fibres in action. We thus see that the 'engines' by which animals move about are fitted with 'dashboards', although, of course, the instruments differ considerably from those of engines.

Not only does the brain know what each muscle is doing, it is also informed automatically whenever the animal's direction of motion is *accidentally* altered. In vertebrate animals this information comes from three small semi-circular tubes or canals lying in the bone near the ears. One canal notifies changes of direction from right to left, another any disturbance of the head upwards or downwards, and the third any tendency to roll from side to side. All this information, together with that coming in through the eyes, is pouring into the brain of the moving animal, and should these messages be cut off, the animal's powers of progression are partially or completely lacking in precision and effectiveness.

Thus each step we take involves not only a very large number of muscular engines but also a very large number of sense organs. And yet we are quite unconscious of the fact that all this is going on; for, in respect of these things, the brain acts mostly as an automatic pilot. Unless something very unforeseen or unusual happens the higher levels of our brain, which are concerned with conscious or voluntary effort, play little or no part in adjusting our normal locomotory movements.

Until a human child is one or two years old, the muscles of his legs are not strong enough to support the weight of his body, and the control mechanism of his brain is not yet fully developed, and so the child cannot use its muscles with sufficient precision to walk. He has to *learn* to walk, just as later on, if he is to swim, he must learn to swim. But in animals this period of helplessness is usually very short, or may not exist at all. The young animal 'teaches' itself; a bird reared in an incubator will fly as soon as its wings have grown feathers enough, and a fish reared from an egg swims perfectly from the day that its tail is fully developed. Even more remarkable, an animal such as a snake or lizard, which has neither seen another animal swim nor ever set eyes on a pond will, when put into the water, swim with full efficiency and grace. Can the explanation of these strange facts be that all terrestrial animals, such as frogs and lizards, are descended from fishes which spent their whole lives in water? The newts, frogs, and toads are, in habit, half-way between fish and lizards; they are still almost equally at home in water or on land. A newt on land moves about in much the same way as a lizard, by means of its legs; but in water the limbs cease to be active and the animal then swims by bending its body like a fish. These things are probably quite automatic: it may be that so long as the newt feels the weight of its body pressing against the ground the mechanism of the legs comes into play whenever the animal moves; but when the weight of the body is carried by the water the swimming pattern is automatically 'switched on'.

2

Swimming

EVERY one of the twelve or more great subdivisions
of the animal kingdom has some representatives that
live in water; only four of these subdivisions have
representatives that walk or creep about on land, and
only two have representatives that fly in the air. These
rather striking facts suggest that animal life began in
water, and that land-living forms evolved at a later
date. This is undoubtedly true in the latest of Nature's
major types – the vertebrates or animals with jointed
backbones. The earliest vertebrate was a fish-like
creature. From the fishes the newts and lizards
evolved, and, later still, the relatives of the lizards
gave rise to the birds and mammals. It is therefore
natural that we should start our survey of animal
movements by looking at those that swim, before we
turn to those that walk or fly.

Man has invented four main types of propellers for
ships; the sail, the jet, the paddle, and the screw. The
sail is different from the others, because the energy
that drives a sailing ship along comes from the wind
and not from engines within the ship itself. This type
of propulsion will not work when the air is perfectly
still, and, even when wind is blowing, the direction
and speed of the ship depend on the direction and

37

speed of the wind itself. Very few animals move by means of a sail; one interesting example is *Physalia*, the so-called 'Portuguese Man-of-War'. *Physalia* is related to our common jelly-fish; it carries a sail or crest and this, projecting above the surface of the sea, functions as a sail. From time to time, numbers of these animals are blown to the shores of England.

No animal propels itself by sending out a continuous jet of water, but some – notably the squids – can dart backwards by emitting sudden squirts of water. The water is sometimes mixed with indian ink and the animal escapes from danger behind a very efficient smoke-screen. It should be noted that the force that moves the squid backwards is exactly equal and opposite to that used by the squid to eject the jet of water from the front of its body.

The sail and the jet are not much used in Nature, but the paddle (in contrast) is widely used. A good example is the turtle, whose limbs are flattened rather like the blade of a canoe paddle. When the paddle is moving backwards, the broad surface of the blade is held square to its line of motion, and thus exerts a powerful pressure against the water; at the end of the driving stroke, the blade is feathered and moves forward, thus exerting much less pressure against the water. Paddles of one form or another are found in a variety of animals (Fig. 12); some have two paddles, some four, and others a larger number, working one after the other in a well-established rhythm.

The paddle may be an improvement on the sail or the jet, but it is far less efficient than the mechanism

a

b

FIG. 12a, b. *The shearwater (Fig. 12a) and the turtle (Fig. 12b), showing how animals swim by means of 'paddles'*

with which Nature has endowed her finest swimmers, fishes, and dolphins. It is with these star performers that we shall be mostly concerned.

The bodies of all fast-swimming fishes, and of dolphins, taper off at each end, rather like a stumpy cigar or an airship. At one end of the body is a large tail or

39

FIG. 13*a*. *The tail-fin of a dolphin or whale is held horizontally and sweeps up and down*

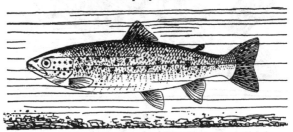

FIG. 13*b*. *The tail-fin of a fish is held vertically and sweeps through the water from side to side*

'caudal fin': the tail-fin of the fish is held vertically, and as the fish moves forward the fin sweeps from side to side; the tail-fin of the dolphin is held horizontally and sweeps up and down, but in both the principles of propulsion are the same (Fig. 13). Clearly, the forward movement of these animals is due to the motion of the tail-fin through the water: the fin must, in some way, press backwards against the water; but how does it do so?

Here at the very beginning of our study, we meet with difficulties. The human eye, as an instrument for exact observation, has its limitations. To watch a

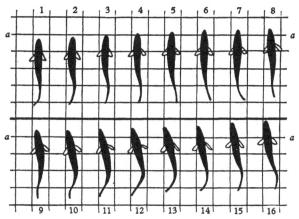

FIG. 14. *When a fish moves its tail from side to side, its body bends under the pressure of the water so that the leading surface of the tail-fin is always held at an angle to the axis of the fish's body, except at the very end of each transverse stroke. In photographs 1–4 the tail-fin is moving from left to right; in photographs 5–7 the slope of the fin is changing; in 7–10 the fin is moving from right to left; in 10–12 it feathers again and then (12–16) again moves from left to right*

moving object with accuracy we must 'follow' the object with our eyes. In other words, each eye has to be constantly on the move, and as we can only reverse the direction of our gaze about twice a second, we cannot watch the tail of a fish all the time unless the animal is moving very slowly. For accurate observation of quick movements we must use a cine-camera. We photograph a fish swimming in a tank at the bottom of which lies a background ruled off in squares. If we know the number of photographs that the camera takes in a second we can find out, not only the frequency with which the tail moves from side to side, but also exactly what the tail does whilst the fish moves forward through a known distance. Photographs taken in this way show that when a fish (such as a trout or mackerel) moves its tail from one side to the other, its body bends under the pressure of the water so that the leading surface of the tail-fin is always held at a slant to the axis of the fish's body, except at the very end of each transverse stroke (Fig. 14). At the end of each transverse stroke the tail-fin reverses its attitude and is thus ready to move towards the other side of the body's axis. The whole motion of the tail is just the same as the motion of a thin sheet of metal (of the proper length, width, and flexibility) immersed in water and made to swing from side to side like a pendulum. These movements are also characteristic of a flexible oar when 'sculled' behind a boat.

As the inclined blade of the fish's tail-fin moves through the water it pushes the water away from its surface in a backward and sideways direction relative to the axis of the fish's body. But the inertia of the water resists this movement, and consequently the body of the fish is acted upon by a force equal but opposite to the force that the tail applies to the water. The amount of water displaced backwards and sideways during each sweep of the tail-fin (and consequently the force applied to the body) depends on the same factors as control the amount of water displaced by the blades of a screw-propeller – namely, (1) the size and shape of the blade or fin, (2) the slant of the blade or fin, and (3) the speed at which the blade or fin is travelling. The fish's tail-fin can be regarded as the blade of a screw-propeller, except that it is travelling from side to side instead of rotating.

The side-to-side movements of the tail of the living fish are due to the shortening and stretching of muscle fibres lying on the right and left sides of the animal's backbone. These fibres appear to be very complicated, for they make the strange zigzag pattern which we see when we remove the skin from the sides of a herring or trout, and the concentric rings characteristic of a steak of cod or salmon. From a mechanical point of view, however, the muscles are relatively simple, for their fibres all run parallel to the sides of the fish's body. When those on the left side shorten, the tail swings over to the left, and when those on the right side shorten, the tail swings to the right. For the muscles to function in this way, the backbone with its joints must act as the fulcrum on which the muscles can rock the hinder end of the body with its caudal fin; but, as we have seen, the muscles cannot turn the tail end of a body without at the same time tending to turn the front end in the opposite way. If the head end of a fish is to travel forward along a straight line at the same time as the hind end of the body (and the tail-fin) is sweeping from left towards right, the front end must be prevented from turning towards the right, and

43

prevented, too, from turning towards the left when the tail sweeps to the left. These counteracting side-to-side movements of the head end are usually very small, because the front end of the body is so much larger than the hind end or tail. A force that is big enough to give quite a large sweep to the tail induces a relatively small movement at the larger front end of the body (Fig. 15).

There is, however, another reason why the front end of the fish does not swing from side to side. It takes a far larger force to move the body sideways through the water, than the force that is required to move it head first forwards. If we like, we could look upon the front part of the fish's body as the part that provides a fixed fulcrum about which the muscles turn the tail and its fin. A system of this sort can be made plainer by a simple mechanical model (Plate 2). The front end of the body is represented by a rod free to slide forward, but prevented from any side-to-side movement by two rails or guides; the tail of the fish is represented by a shorter rod hinged to the first one, and in contact at its other end with a fixed slanted bar. If the rods are connected by a stretched spring, the front rod glides forward between its guides whilst the far end of the hinder rod slides along the inclined bar. Plate 1 shows that a living fish can in fact propel itself forward by exerting its effort against fixed pegs arranged in the same pattern as the rigid guides and inclined track of the model. The muscles of the fish brace the front part of the body against the water just as they brace it against the rigid pegs. The tail of the fish pushes backwards and sideways; the sideways push causes little or

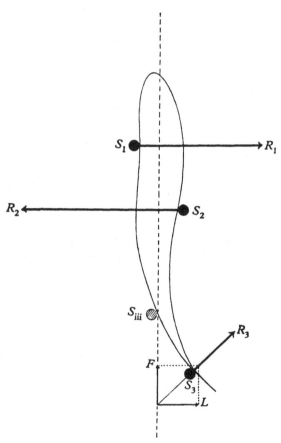

FIG. 15. *Diagram of Plate Ia. The head of the fish is prevented from turning by the pegs* S_1 *and* S_2, *while the tail of the fish is pushing backwards and sideways against* S_3. *The force* F *driving the fish forwards is the forward component of the force* R_3 *exerted by the peg* S_3 *against the tail. The side force* L *exerted against the tail by* S_3 *is neutralized by forces exerted by* S_1 *and* S_2

45

no movement because it too is opposed by the pegs or water pressure operating against the front end of the body. Or again, we can visualize the motions of a fish by forcing the tail of a rubber model to move in the same way as the tail of a living fish. This can be done by means of a wave machine. The backward drive of the model's tail can be observed by watching the movement of particles suspended in the water.

The smooth forward motion of a fish such as a trout or mackerel depends on the existence of a broad tail-fin. Yet some fish, notably the eel, can swim quite well although they have no well-developed tail-fin. At first sight, the wriggling motion of an eel seems to be very different from the sort of movement we have been talking about; but in fact the difference is much more apparent than real: it is almost entirely due to the difference in length and flexibility of the body. In an eel, the body is long enough to allow one part to be moving towards the right whilst another is moving towards the left, and it is this effect that produces the wave-like movements seen in the swimming eel. As with a trout, an eel can propel itself forward only when it exerts its movements against the external world. On a smooth level board, an eel out of water wriggles in vain; but as soon as it is put on a board studded with smooth pegs, it glides forward with remarkable smoothness and speed (Fig. 16): it can also glide rapidly through a 'wavy' glass tube.

When a fish at rest in the water starts to swim, at first the force exerted by its tail is used to overcome the inertia of the body, but as it begins to move forward in

46

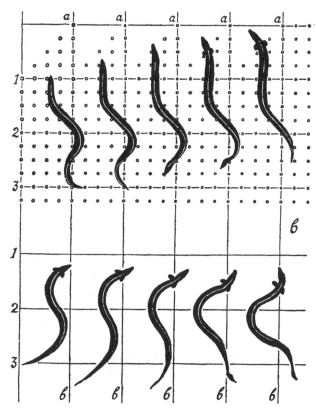

FIG. 16. *An eel which can glide forward over a board studded with smooth pegs* (*Fig. 16a*) *cannot progress over a board without pegs, though its body may develop very large muscular waves* (*Fig. 16b*)

the water it meets with a resistance due to friction between the water and the whole surface of the fish moving through it. The faster the fish moves the

47

greater this friction grows, until a time comes when the whole effort of the muscles is being used up in overcoming this water resistance or friction, and none to spare for making the fish swim faster; the fish swims at a steady speed when the driving thrust of its tail is equal but opposite to the drag of the water flowing past the body. The limit of this steady speed depends on the resistance of the body to the surrounding water. A study of models of different shapes shows that water can flow most readily over a body shaped rather like a blunt-ended cigar. Objects of this shape are said to be *stream-lined*, since water streams over their surface along well-defined lines and is not broken up into eddies and similar disturbances. The shape of the body of a mackerel or a herring is very close to a stream-lined form. The proportions of the body are right, and the whole surface is smooth; it is free from projections and the head merges smoothly into the trunk and the trunk into the tail: in fact the bodies of these fishes are beautifully adapted for being driven through the water with a minimum of effort.

When we walk along the river-side and catch sight of a trout darting through the water, we get the impression of very high speed. Different people make different guesses, but most anglers are inclined to think that 15–20 m.p.h. is not unreasonable. We tried to check these guesses in the laboratory – with rather surprising results. A small trout (about 9 in. long) was put into a tank and allowed to settle down. Then a cine-camera above the tank was started up and the trout suddenly startled by a moving shadow. The fish moved at what appeared to the eye to be a high speed, but when the photographs were developed and analysed the greatest speed was found to be no more than 7 ft per sec., or 5 m.p.h. Here again it was found that the eye is a poor instrument for scientific observation, for it proved to be quite unable to follow a movement occurring in an unexpected direction at an unexpected moment. So far we have not been able to make reliable measurements of the maximum speeds of larger fish, but such experimental results as we have, suggest that 10 m.p.h. would be somewhere near the greatest speed that a 20 lb. salmon could keep up for a period such as 20 sec. To keep up a speed of 5 m.p.h., the 9 in. trout would have to exert a backward thrust with its tail of about one-half to one-third of its own weight, and a similar effort would be needed from a salmon travelling at 10 m.p.h. Nothing very impressive, you might say. On the other hand, the photographs of the trout revealed one very interesting fact, for they show that the fish gets up to its maximum speed within $\frac{1}{20}$ sec. starting from rest; this is an acceleration of 140 ft per sec. in one second, and could only be produced by a thrust equal to at least four times the animal's own weight – since a force *equal* to the animal's weight would produce an acceleration of 32 ft per sec. per sec. All these observations suggest that a fish can exert very powerful forces, but only for very short periods of time – they can in fact 'jump' through the water but they cannot keep up high speeds for more than a short time.

But if a fish does not usually exert a force of much

more than one-third of its own weight, how are we to account for the photograph shown in Plate 4, where a swordfish has driven itself through the planking of quite a strong boat? Still more, how are we to account for the fact that 'swords' have been found driven nearly 2 ft into the thick timbers of ships? A little reflection will show, however, that the energy set free when the swordfish strikes the side of the dinghy, does not come from the movements of the fish's tail but from what is called the kinetic energy stored in the animal's whole body. If a fish, weighing 600 lb. and travelling at 10 m.p.h. runs into the side of a boat and is thereby brought to rest in a distance of 3 ft, the average force applied to the boat is about one-third of a ton, and the whole of this force is applied over an area – the tip of the sword – of about one square inch. The blow would be the same as the blow of a sledge hammer weighing 10 lb. and meeting the boat at a speed of about 80 m.p.h. No wonder then that the side of the boat was shattered! And, if a swordfish of 600 lb. travelling at 10 m.p.h. meets, end-on, a wooden ship travelling at 10 m.p.h. in the opposite direction, the average force applied at the point of the sword is about 4½ tons. The surprising thing is, not that the sword penetrates the wood, but that it does so without itself being smashed in the process.

Before we leave the speed and power of fish movement, let us look for a moment at the mechanics of angling (Fig. 17). When a fish takes the fly, and begins to run out the line from a reel, we have the impression of high speed and great strength; but we must bear in mind two things. First, each revolution of a trout-reel seldom pays out more than 1 ft of line, so that if the reel is revolving 4 times per sec. (thus creating a good deal of excitement!) the speed of the fish's movement cannot be greater than 4 ft per sec. (about 3 m.p.h.). Secondly, the fish is exerting its pull at the end of a long lever. If the trout exerts a pull of ½ lb. at the end

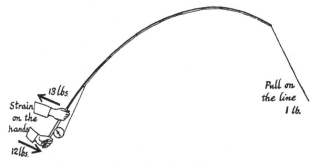

FIG. 17. *Though the pull exerted by a large fish is only 1 lb., each of our hands must sustain a force of 12 or 13 lb. at the other end of a salmon rod*

of a 10 ft rod, the rod can only be held stationary if, with our hands 1 ft apart, each of our arms sustains a force of 4–5 lb. In other words, the sound that a running reel makes, and the forces that the rod exerts against our hands, may give a very misleading impression of the speed and power of the fish's movement. A light trout rod cannot, safely, be trusted to sustain a pull of more than about 4 oz.

Do not suppose that an accurate knowledge of fish speeds, interesting though it is, is of no use. All our salmon fisheries depend on the power of the full-grown fish to migrate upstream on the larger rivers, and so to

51

come to smaller streams, where they spawn. Many
such rivers are obstructed by waterfalls or by places
where the stream is flowing more than 10 m.p.h. To
help them in this difficult up-stream migration, 'fish-
passes' are built, where the flow is reduced or where
resting-places are provided for fish which have tired
themselves in their efforts to overcome the pressure of
rapidly flowing water. The leaping of salmon in the
neighbourhood of falls is, of course, a familiar sight,
and no doubt some fish do really get up the falls in this
way (Plate 3). But it is to be doubted if this is the normal method of ascending a fall; it is more likely that
most fish actually swim up in the sheet of water that is
flowing over the fall. A sudden impulsive effort, sufficient to lift the fish 6 ft into the air above the surface of
the water, would, if applied under water, give a forward speed sufficient to drive the fish up the fall without leaping out into the air. All our knowledge seems
to show that fish can develop powerful propulsive
forces but only for a very short time; their muscles tire
easily and they quickly get 'out of breath'. But we
shall return to this in a later chapter.

If I were asked to name Nature's finest swimmer I
should undoubtedly choose the dolphin. The dolphin
can maintain a speed of 20–25 m.p.h., and although
we do not know precisely how long this speed can be
kept up, the dolphin certainly does not tire anything
like so quickly as a fish. Dolphins are mammals; they
are warm-blooded, and they breathe with lungs as we
do. These advantages enable the animal to renew its
muscular energy much faster than a cold-blooded fish

can. Still, even allowing for these improvements in design, the dolphin is a very remarkable animal: if the resistance encountered by a dolphin gliding through water is the same as that of a rigid model of the same shape and size, the muscles that move its tail up and down must be about ten times more efficient than those of a dog or a man. Alternatively, if the power of

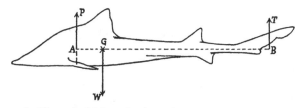

FIG. 18. *When a dogfish is moving forward through the water, the horizontal pectoral fins produce an upward force* P, *acting through a point* A *on the horizontal axis of the fish, whilst the vertical component* T *of the force exerted by the tail acts through* B. *The point* G *through which the weight of the fish* W *acts, is the centre of gravity, and the moments of* P *and* T *about* G *must be equal and opposite if the fish is to maintain an even keel. The vertical fins of the fish act like the vertical fins at the end of an arrow and prevent the fish's body from yawing from side to side*

the muscles of a dolphin is the same as that of the same weight of muscle in other mammals, then the water must flow over the dolphin's body with ten times less disturbance than over a rigid model. If this is so, Nature's design for a dolphin is much more efficient than any submarine or torpedo yet produced by man.

One most striking fact about a fish in movement is its ability to move on an even keel and to change its direction of movement rapidly without losing balance. Comparing a fish gliding through water with an arrow travelling through air, one might well guess that the

smoothness of the fish's motion is controlled by the fins
projecting from the body, just as the motion of the
arrow is controlled by the feather fins attached to the
end of the shaft. In primitive fishes (as in the modern
dogfishes and sharks), the fins are arranged in two
series – 'vertical' fins project upwards and downwards
from the back and the lower surface of the tail, and
paired 'horizontal' fins situated at the shoulders of the
fish and just in front of the muscular tail (Fig. 18). By
observation of a carefully prepared model suspended
in a current of air, Professor J. E. Harris has shown
that the unpaired 'vertical' fins, like the vertical fins
of an arrow, are largely concerned with the correction
of any tendency of the fish to 'yaw' from side to side or
to 'roll' about the long axis of the body; and that the
'horizontal' fins are related to the control of 'pitching'
movements in which the head tends to tip upwards or
downwards. In sharks and dogfishes, these paired fins
act as flat plates (or hydrofoils) whose angle of inclination to the moving water can be varied by muscular
action. If the head of a fish is accidentally pointed
downwards by a current of water, the 'set' of the fins
changes in such a way as to tilt the head upwards
again. Similarly, if a minnow is forcibly rolled to one
side, all the fins work together to bring the fish back to
an even keel.

This righting effect of the fins is largely automatic,
and is due to the existence of three semicircular tubes
or canals inside the head of the animal. If the body
yaws, pitches, or rolls accidentally, one or other of
these three canals sets up impulses in the nerve leading

from the canal to the brain, and the brain at once sends appropriate motor impulses to the muscles moving the fins. When a fish swims into a current of water, as the head of the fish enters the current it begins to yaw downstream, but this turning movement at once affects the horizontal canal of the ear, and impulses pass down the nerves to the muscles of the body, which promptly turn the head of the fish upstream again.

We must not suppose that a fish relies solely on its semicircular canals for controlling the accuracy of its movements. These sense organs only come into use when the head of the animal is pitching, rolling, or yawing. When a fish wishes to maintain its direction along a straight line – as when a pike darts at its prey or when a fish 'keeps station' in a perfectly straight current of water – it relies largely on its eyes.

In addition, all fish are equipped with sense organs able to respond to very small disturbances in the water near their bodies. These can often be seen on the head and sides of a fish as a series of minute openings, which connect with a tube or lateral line canal. Within this canal are bunches of very fine hairs – rather like cilia – which send nervous impulses into the brain of the fish whenever the outside water is disturbed. You may have noticed how difficult it is to push a net close under the body of a fish, even though the fish cannot see the net. This is because of the lateral line system. Again, we find a strange blind fish (*Anoptichthys*) which, nevertheless, is able to feed on small organisms because it can detect them by the disturbances they make in the

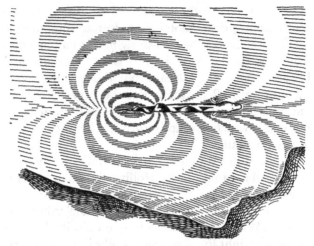

FIG. 19. Gymnarchus, *a fish living in muddy African rivers where its eyes are of limited value, surrounds itself by an electric field, whose general nature is indicated by the curved bands in the figure. The fish is warned of the approach of a strange fish or object by a change in this electric field*

water by their own movements. Perhaps the most remarkable way by which a fish knows what is happening in its neighbourhood is that recently discovered by Dr Lissmann. As you know, when the 'electric eel' (*Electrophorus*) is excited it produces a powerful electrical discharge – strong enough to be made to light up an electric lamp. Dr Lissmann has now discovered that this method of defence is only a special case of a much more exciting state of affairs. In the warm, muddy rivers of the Gold Coast there are fish (*Gymnarchus*) which are constantly sending out a succession of small electric impulses, surrounding themselves in this way by an electric field (Fig. 19); whenever a strange fish or object comes near the fish, the electric field is altered, and in some way, not yet understood, the *Gymnarchus* becomes aware of this. These fish have indeed something like a radar equipment of their own, which largely replaces the eyes as a means of knowing what is happening in the surrounding muddy water.

If you watch the fishes in an aquarium, such as that at the London Zoo, you will very soon discover for

FIG. 20. *The mudskipper* (Periophthalmus) *can live out of water for quite a long time. It supports its body on its pectoral fins, and so shows how the paired fins of a fish later evolved into supporting props or limbs*

yourself that what I have told you about the swimming of a fish is only a part of the whole story. You will easily find fishes which do not use their tails as propellers, but rely on their fins alone. Sometimes, as in *Gymnarchus*, the fin on the back of the fish is almost the sole means of propulsion. In other kinds of fishes, such as the skate, the breast-fins are enormously enlarged, and replace the tail as organs of locomotion.

All these modifications of the fins are interesting,

57

but, as we shall see in the next chapter, we shall have to pay particular attention to that type of fin modification which is characteristic of fishes that often remain at rest on the bed of the water. Such fishes often keep the tips of their paired fins in contact with the ground, using them as props to stop themselves from rolling over. A few, the mudskipper (*Periophthalmus*) for example (Fig. 20), go further than this, they can even creep out of water and support the front of their body on their breast-fins, much as a lizard supports its head and shoulders by its front limbs; they can also propel themselves on land for quite a long way by hopping along on their tails and fins. This is not to say that land-living vertebrates, such as newts and lizards, are descended from the mudskipper, but you can take from this curious little fish a useful hint about the way in which Nature, starting with a fin which was first of all a 'balancing' organ, by endowing it with strength and power to move, first changed it into an extra propeller and finally transformed it into a jointed limb, capable of supporting the weight of the body and of moving the body forward in a way characteristic of a land-living newt.

3

Walking and Running

WE have seen how a typical fish moves itself along by means of a tail, and balances itself by means of fins. Now we shall look at land-living animals – animals that hold their bodies off the ground and move themselves along by *limbs*. At first, the contrast between a fish and a horse seems so great that it is difficult to imagine how one could have been derived from the other by any natural process. The problem that Nature had to solve is something like that facing an engineer who is asked to produce a car from a submarine without ever taking it to pieces completely or putting it out of commission. Let us try to see some of the steps by which Nature achieved success.

We cannot run till we have learnt to walk, and we cannot walk till we are able to stand; therefore the first step towards the transition into a land-living animal from a fish was taken when the animal was able to stand – to balance its body against the pull of gravity. So long as a fish is under water the downward pull of gravity is balanced by the upward buoyancy of the surrounding water; but, in air, if we try to balance the body of a fish, with the lower part of its deep and rather narrow body resting on the ground, we find that it falls over on to its side. It falls over because it is impossible for us to ensure that a vertical line drawn downwards from its centre of gravity shall always fall inside the area of contact of the body with the ground.

Quite a number of fish get over this difficulty by spreading out their breast-fins sideways from the body and keeping the ends of the fins on the ground. The fins then act as props, and the centre of gravity of the body can readily be kept over the triangle marked out by the tips of the fins and the point at which the hind end of the body is in contact with the earth; we have already mentioned the mudskipper (*Periophthalmus*) (p. 58 and Fig. 20) as an example. To serve as props, the fins must be fairly stiff, and tightly braced to the lower surface of the body. The paired fins of most fishes have these two features anyway, for their fins are stiffened by a bony skeleton, and are tied on to the body by the muscles which, during normal aquatic life, move the fins up and down or to and fro.

The next stage in the development of a standing animal was reached when both pairs of lateral fins (breast-fins and pelvic-fins) were used as lateral props and were strong enough to allow the whole body to be held above the ground. No known land-living fish uses its fins in this way, but the lung fishes (which are nearest living relatives to the land-living newts) often rest with the tips of their fins touching the bed of the streams in which they live. A reconstruction of an early land-living fish is shown in Fig. 21 (3), where it will be seen that the ends of the fins in contact with the ground are flattened and so spread the weight of the animal over a surface large enough to prevent the fins

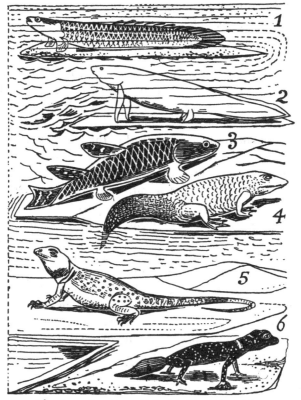

FIG. 21. *The change from fins to limbs.* (*1*) *The African fish* Polypterus *supports the front end of its body by its pectoral fins.* (*2*) *A similar use of the fins is found in the lung fishes – which are probably the nearest living relatives of land-living vertebrates.* (*3*) *A reconstruction of a fossil fish* (Eusthenopteron) *which, for various reasons, is known to be a close relative of the earliest land-living vertebrates: note how the body is supported by both pairs of horizontal fins.* (*4*) *An early land-living vertebrate* (Diplovertebron): *note how the elbows and knees are held 'akimbo'.* (*5*) *A lizard at rest, supporting itself in a way very similar to* Polypterus. (*6*) *A lizard* (Gymnodactylus) *standing with the whole of its body off the ground*

sinking into a soft muddy surface. An animal like this could balance itself easily, because its centre of gravity would lie well within the oblong marked out by the four lateral props or limbs.

The internal structure of fish fins varies a good deal in different kinds of fish, but when fins began to be used as supports for land-living animals the internal skeleton became very similar to that of a human arm or leg. The limb was attached to the body by a single bone; this bone was hinged at its outer end to two other long bones (forming the forearm or shank), and these in turn were united to a flattened hand or foot by a series of small bones forming the wrist or ankle. As there are five fingers on each of our hands, the basic plan of the limbs of vertebrate animals is spoken of as 'pentadactyl'. At present, for us here, a detailed study of the anatomy of these limbs is not very important, but we ought to remember three main facts: first, that the existence of 'elbow' and 'knee' joints enables an animal to bend the limb and so lift its feet off the ground; second, that the ability of the limb to act as a rigid support for the body depends on the presence of muscles; and third, that for the limb to do its work at its best, more depends on the form of its hands and feet than on its upper parts. Let us look at the third of these facts first.

An animal standing on its four legs can be compared with a table. If we wish to load a table when it is standing on soft mud or sand, the feet must have large bearing-surfaces, comparable with the webbed feet of a frog or duck; but if the table (or animal) is standing

62

on hard ground, the ends of the legs can be hard and
can have relatively small bearing-surfaces, such as
those of a horse or a sheep. Neither the webbed feet of
a frog nor the hooves of a horse are particularly well
suited for standing on steep slopes, and it is for such
conditions that the feet of many vertebrate animals
are specially adapted; such feet have claws, suckers
(tree frogs), or fingers (e.g. chameleons, monkeys)
which can clasp branches of trees or other steeply inclined objects. Extreme cases of such adaptation are
found in sloths, which make a habit of hanging upside
down by their long, curved claws, and in the geckos,
where the sole of the foot can be pressed so close
against a smooth surface that a gecko can hang from
the underside of a level sheet of glass.

All these kinds of feet, however different in form
they may be, are operated by the two upper parts of
the limb (the upper arm and forearm, or the thigh and
shank) which are remarkably alike throughout the
whole range of four-legged animals. Within this range
we find, however, a very marked difference in the
length of the leg: a horse has, for its size, much longer
legs than a rat or a lizard. There are, no doubt, many
advantages in this, but one of these becomes obvious
when models of such animals are placed on a surface
that can be tilted up. A plumb line is attached to the
lower surface of the model immediately beneath its
centre of gravity. As soon as the model of the horse,
being tilted, reaches an inclination of about 30°, the
plumb line begins to fall behind the hindfeet, and the
model tumbles over backwards (Fig. 22); whereas

with the model of a rat or lizard this critical stage is not reached until the supporting surface has become almost vertical. We can see that the short legs and long bodies of a rat or squirrel are adapted for life on steep

FIG. 22. *As soon as this model of a horse is tilted to an angle of about 30° the plumb-line from the centre of gravity falls behind the hindfeet and the model tumbles over backwards*

slopes or on branches of trees, whereas – as we shall see later – the long legs and shorter bodies of horses are adapted for life on relatively flat surfaces.

In all the earlier types of four-legged animals (newts and lizards), the limbs are held 'akimbo', with the elbows and knees well away from the sides of the body. In general, these animals only lift their bodies off the

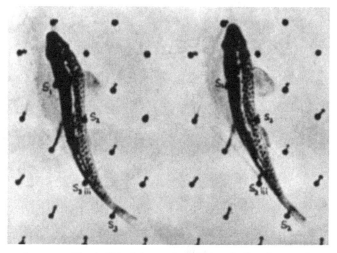

1. *A trout propelling itself in air by pressing its body and tail against smooth pegs (S1-S3) projecting from the surface of a board.*

2. *A model fish in which the muscles are replaced by a spring which presses the body against a rod which can slide forward on smooth bearings and presses the tail against a curved strip of metal. When the spring comes into action the fish slides forward.*

3. *A salmon sometimes ascends a waterfall by jumping into the air, but – more often – it swims up in the sheet of water flowing over the fall.*

4. *A swordfish which has driven its head through the side of a boat. When a fish weighing 600 lb. and travelling at 10 m.p.h. is brought to rest in a distance of 3 ft, the average force applied to the boat is about one-third of a ton.*

5. *When a man is running, his foot, like that of the galloping horse, is travelling backwards quite rapidly before it is put down almost vertically below the hip joint. Note also how the leg is flexed whilst swinging forward and how the left arm swings forward with the right leg (and vice versa). All these features facilitate efficient running.*

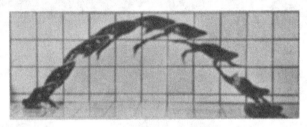

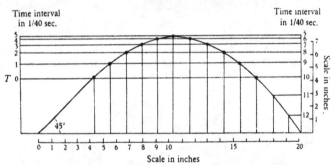

Time interval
in 1/40 sec.

Time interval
in 1/40 sec.

Scale in inches

Scale in inches

6. *Photographs of a jumping frog. The top figure shows the hindlimbs extending at an angle of about 45° to the vertical (time interval $\frac{1}{40}$ sec.). The lower photograph shows the trajectory of the body through the air: the length of the jump was about 20 in. (time interval $\frac{1}{40}$ sec.). The diagram shows the position of the centre of gravity of the body at intervals of $\frac{1}{40}$ seconds after the take-off at T.*

7. *Just before a take-off all the joints of the hindlimbs of a locust or grasshopper are tightly folded up at the sides of the body; as soon as the jump begins these joints extend. The limbs extend to their maximum extent in about $\frac{1}{30}$ second.*

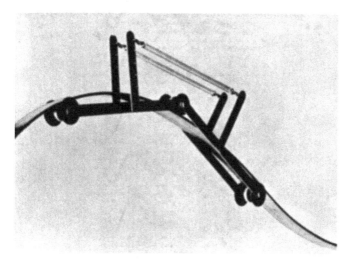

8. *Model of a snake's movement. Two rigid rods hinged together move forward along a curved track when a spring shortens across the hinge uniting the rods and when the track is so shaped that its curvature alters along its length.*

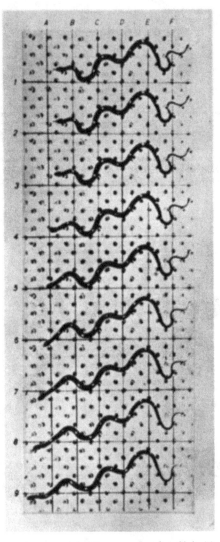

9. *A grass-snake gliding over a board studded with smooth pegs. The animal pushes backwards and sideways against the pegs just as an eel presses backwards and sideways against water.*

10. *Flying fish. The upper figure shows the fish 'taking-off' by beating its tail against the water. The lower figure shows the fish in full flight.*

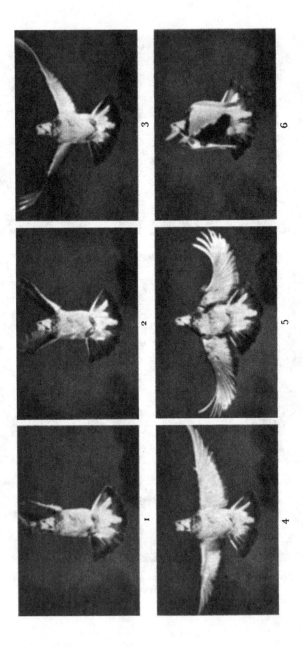

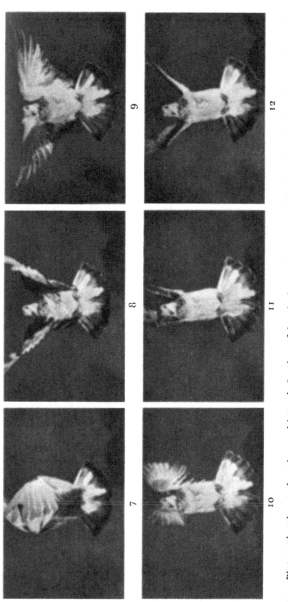

11. *Photographs taken at about* $\frac{1}{80}$ *second interval of a pigeon flying slowly towards a camera. Note how the wings beat downwards and then forwards until they meet in front of the bird's head; they then strike backwards and upwards.*

12. *Similar photographs of a pigeon flying slowly across the field of the camera. Note how the wrists are raised at the beginning of the up stroke and how the primary feathers then strike rapidly backwards and upwards.*

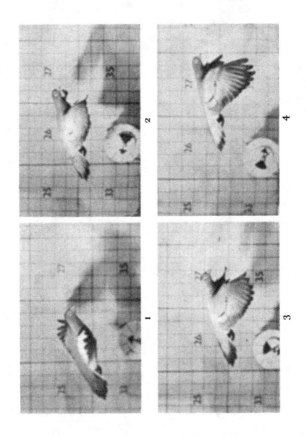

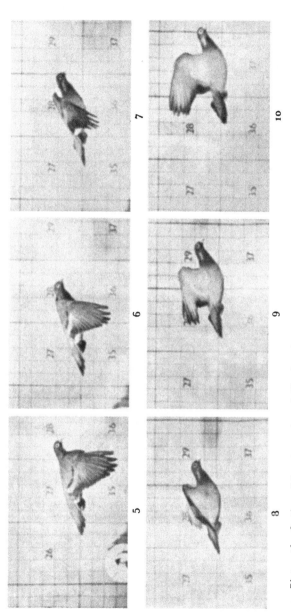

13. *Photographs of a pigeon flying at normal speed. Note how the wings no longer swing forward in front of the bird's head and how the backward flick at the end of the up stroke is greatly reduced.*

14. *A gannet in gliding flight.*

15. *The flying squirrel (Glaucomys) jumps off trees and, spreading its arms and legs, glides downwards gracefully for about fifty yards.*

16. *Photograph of a bat in gliding flight.*

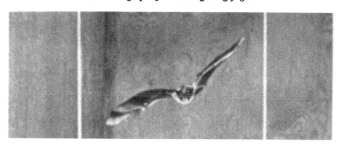

17. *A bat can avoid obstacles in the dark. The photograph was taken just as the animal was passing between two vertical wires almost as close together as the width of the animal's wing span.*

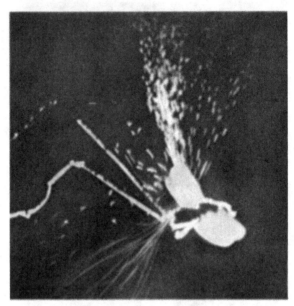

18. *In order to demonstrate the propulsive and lifting effect of a fly's wings, Dr Hollick photographed the insect when its wings were beating in a shower of very fine particles. In the developed photograph, the particles appeared as lines whose length showed the distance travelled by the particles during the time of exposure of the photograph. Notice how the particles are drawn slowly towards the wing region and then projected very rapidly downwards and backwards. The reaction from this downward and backward current of air drives the fly upwards and forwards.*

ground when they are going to walk; if they are at rest, a good deal of the weight of the body rests directly on the ground. But in later types (mammals) the elbows and knees are drawn in and lie vertically under the joints (shoulder and hip) by which the limbs are joined to the body.

With animals who stand up in this fashion the comparison with a table becomes still closer, and it is useful to consider what happens if we put a heavy weight on the top of a table and then attempt to take away one of the four legs (Fig. 23). If the weight is in the middle of the table, the table will topple over if any one of its four legs is removed. If, however, the weight is near to one end of the table, either one of the two legs at the other end can be removed without upsetting the table; collapse comes when we try to take away one of the two legs that are near to the weight. We can see, then, that so long as an animal's centre of gravity does not lie exactly midway between its front- and hindfeet, the animal always has two alternative triangles of support – triangles made by the two feet that are nearer to the centre of gravity and *one or other* of the two that are farther away from it.

Throughout the great group of mammals, we find there are two main classes of animals: those where the centre of gravity is nearer to the forefeet, and those where it is nearer to the hindfeet. The horse is an example of the first type. You will have noticed how a horse, standing quietly in a field, often stands with one hindleg resting quite gently on the ground (Fig. 24); a horse, too, can kick out with a hindleg without falling

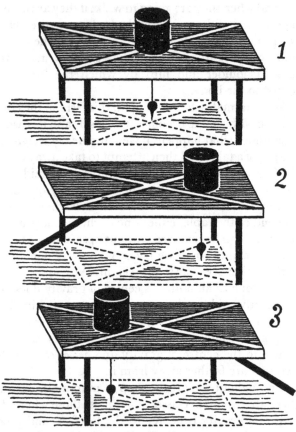

FIG. 23. *A four-legged animal can be regarded as a table with a weight resting on its top. Except when the weight lies at the centre of the table (Fig. 23, 1), one leg of the table can always be removed without causing the table to topple over. The condition of stability depends on a plumb-line from the centre of gravity meeting the ground inside the triangle formed by the three legs which remain on the ground*

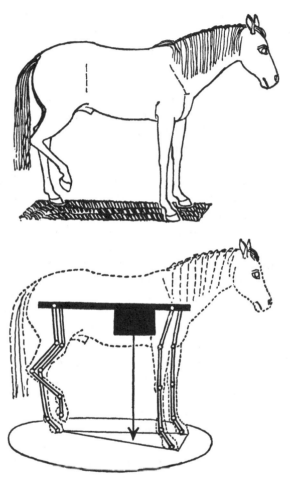

FIG. 24. *The centre of gravity of a horse's body normally lies nearer to the forefeet than to the hindfeet – consequently it can stand with one hindfoot off the ground*

over. But you can only raise the *front* foot of a horse by forcing the animal to shift its weight on to both hindfeet (Fig. 25). We shall see later that the habit of keeping the weight well forward towards the frontfeet is an adaptation for rapid movement.

Animals such as rabbits, squirrels, and bears belong to the other class. All these stand with the weight normally well back towards the hindfeet, and consequently either frontlimb can be lifted (Fig. 26). Indeed, within this second group, Nature has allowed some animals to bring the centre of gravity of their bodies so far back that it lies above the hindfeet themselves; no weight at all then falls on the frontfeet and the animal becomes two-footed, or bipedal. This is possible only when the bearing-surfaces of the feet form a fairly large area of contact directly under the line of action of the animal's weight. And as for the kangaroo – his centre of gravity lies *behind* the hindfeet, and the triangle of support for him is provided by the two hindfeet and the tail; as Mr Lane puts it 'the kangaroo carries its own shooting-stick!' (Fig. 27). We must, however, now return to more normal animals, and try to understand how they use their limbs as propellers.

When a bear or a horse is walking, each limb performs a well-defined series of movements. The foot is placed on the ground a little in front of the shoulder, or the hip joint, and remains on the ground until the shoulder or hip joint has been moved forward and is in front of the foot; then the foot is lifted and swung freely forward to be ready for the next step. The movement

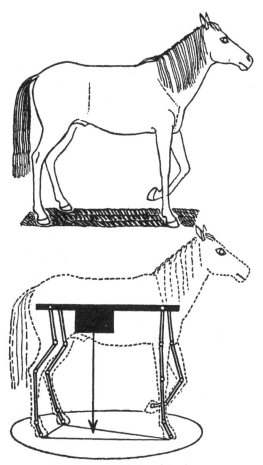

FIG. 25. *If a horse is to lift a forefoot, its body must be moved backwards until the centre of gravity lies within the triangle marked out by the two hindfeet and the forefoot which is not being lifted*

of a leg relative to its upper joint is controlled by the muscles that join the leg to the body. The muscles that pull a leg backwards while it is moving the body forwards are called the *retractor muscles*; those that swing the leg forward in the air are called the *protractor*

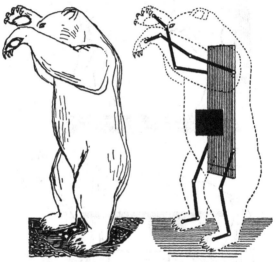

FIG. 26. *A bear can bring the centre of gravity of its body to a point vertically over the area marked out by the soles of its hindfeet and thus become bipedal*

muscles. The complete cycle of movement is best followed if we start at the point at which a hindleg is lifted off the ground and begins to swing forward in the air. At this moment the knee is bent, and even if all the muscles of the hip are quite limp, the leg will begin to swing forward under its own weight, like a pendulum; the force of gravity accelerates the foot downwards, and the limb's inertia causes it to swing past the point at which the foot is vertically under the hip.

If an animal wants to swing its leg forward faster or farther than it would travel under its own weight, the

FIG. 27. 'A kangaroo carries its own shooting-stick' (Lane); it can tilt its body backwards beyond the vertical and support its body on its two hindfeet and tail

protractor muscles must be brought into action: if the animal wants to swing a leg slower than the swing that would be produced by the limb's own weight, the re-tractor muscles must come into action and act as a brake. The speed with which any particular limb can swing forward under its own weight depends upon the length and shape of the limb, and this speed is obviously the one that requires least muscular effort. As soon as the leg has reached its forward position, the knee joint extends, the foot is placed on the ground, and the leg begins to carry some of the weight of the body. At this point the leg begins to behave as the spoke of a wheel, for it is being turned by the retractor muscles of the hip in essentially the same way as the spoke of a wheel turned by its axle. If the foot were not firmly fixed to the ground, the leg would slip backwards, just as the driving wheel of a car skids (spinning backwards) on slippery ground. If the foot holds firm on the ground, however, the turning effect of the retractor muscles drives the body forward, because the foot is acting as a *fixed* fulcrum. It is important to note that the practical effect of the retractor muscles depends entirely on whether the foot is firmly fixed to the ground or not: if the retractor muscles of the hindleg of a horse pull on the leg when the foot is off the ground, the horse 'kicks' backwards, whereas if the foot grips the ground, the same muscles propel the body forwards. Many of these facts can be illustrated by means of a model (Fig. 28).

When an animal is walking slowly, the retractor muscles provide the main driving force until the foot is vertically under the hip; but from that point onwards another principle comes into play, for the knee joint gradually straightens out during the second half of the step. This straightening process pushes the body forwards and upwards, without any help from the muscles of the hip; the effect is the same as that of an extensible pole or strut.

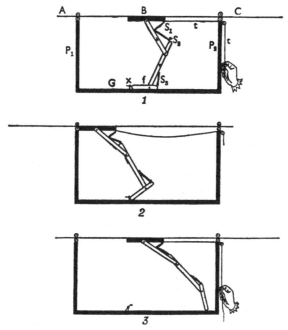

FIG. 28. *Model of a limb. The body of the animal is represented by a long rigid rod* ABC *supported by two pillars* P_1 *and* P_2. *The limb (consisting of three jointed segments* (h, r, *and* f)) *is attached to the body by a hinge. Across this hinge and each joint is a spring* (S_1, S_2, S_3). *The length of these springs is so adjusted that they are stretched when the body is pulled into the position shown in Fig. 1 by means of the string* t.

If the string is released when the foot (f) *is fixed to the ground by a hook* (x), *the body slides forward as in Fig. 2, the joints extend, and the heel rises from the ground – just as when a bear propels itself with a front limb.*

If, whilst still holding the string t (*as in Fig. 1*), *the hook* (x) *is released, the limb swings backward as in Fig. 3, and the body does not move forward. In order to propel the body, the foot must be kept fixed to the ground*

73

The hip muscles operate the limb as a man might operate a canoe paddle with its end against a fixed point on the bottom of a stream; but the extensor muscles of the knee operate the limb as a man uses a punt pole.

As the forces that a limb applies to the body are exactly equal but opposite to those that the foot applies to the ground, we can measure them by allowing the foot to make its thrust against a platform suspended on suitable springs, one group of springs responding to vertical forces and another group to horizontal forces (see Fig. 3, p. 16).

Most of the larger land-living animals have four limbs. Insects have six, spiders eight, and millipedes hundreds. How do all these limbs work together to produce smooth and steady progression? Of four-legged animals we can see that only one leg can be off the ground at any instant, if the animal is not to fall over. A foreleg can be lifted if the centre of gravity of the body is far back enough to ensure that the weight of the body is carried by both hindfeet and one forefoot; a hindfoot can be lifted if the centre of gravity is far enough forward. In short, a leg can be lifted from the ground and swung forward provided it is not one of those which are essential for carrying the weight of the body.

How is this condition satisfied? If you watch any four-legged animal walking very slowly but steadily, no matter whether it be a newt, a toad, a tortoise, a chameleon, a sheep, or an elephant – or even a young child on all fours – you will find that the pattern of

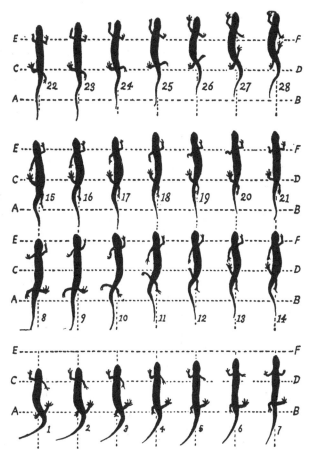

FIG. 29. *Diagonal sequence of limb movements in a newt. In positions 1–7 the right forefoot is being moved forwards, whilst the other three feet are on the ground. In positions 8–14 the left hindfoot is moving forward; in positions 15–21 the left forefoot is moving forward and in positions 22–28 the right hindfoot. When the animal is walking slowly there are always three feet on the ground, whilst the fourth is swinging forward*

75

limb movements is always the same (Fig. 29). In every case the four legs are lifted from the ground (or replaced on the ground) in a definite order; if we begin to watch when the right forefoot lifts (Fig. 29, 1–7), the next leg to lift is always the hindfoot on the left side (8–14); this is followed by the left forefoot (15–21), and this in turn by the right hindfoot (22–28); after this the right forefoot lifts again and the order is repeated. When we work out the geometry of this diagonal pattern, we find that it is the only order of stepping which conforms to the requirement that no foot should ever be lifted unless the centre of gravity of the body lies over the triangle marked out by the other three. As each foot comes down it forms the corner of a new triangle of support, and as soon as the centre of gravity comes to lie within this triangle the fourth foot, the one not involved in this triangle, can move. In other words, four-legged animals, when moving slowly, move their legs in such a way that it is possible for the body to stop at any instant without falling over.

If three feet are always on the ground whilst the fourth is swinging forward in the air, each complete cycle of a limb's movement can be divided into four units of time; during three of these units, the foot is on the ground driving the body forward and supporting some of its weight. The total distance through which the body moves forward during these three periods is determined by the length of the limb, and the angle (relative to the hip or shoulder) through which it turns between the moment when the foot is placed on the ground and the moment when it is lifted. Call this distance the 'span'; it follows then that the distance between two successive footprints of the same foot (known as the 'stride') must be 1⅓ spans, because the body is moving forward one-third of a span during the time when each limb is being swung forward. If the span were 3 in., the stride would be 4 in., and if the

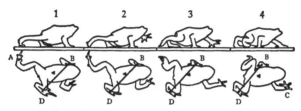

FIG. 30. *Outlines of four successive frames of a toad in normal fast walking, showing the support of the body at different stages of the stride. The side views are mirror views corresponding to the dorsal views. The black triangle represents the position of the centre of gravity of the body; the black circles on the feet represent stages in which they are bearing some of the weight. The centres of pressure of the feet B and D are joined by a line*

animal moves each foot once a second, the speed of advance is 400 yd (less than a quarter of a mile) per hour.

Very few animals are content to move at this rate – the speed of a sedate walk; they quicken their movements by sacrificing the ability to stop at any instant without loss of balance. As the rate of walking gets faster (Fig. 30), each foot is lifted off the ground *before* the one next ahead of it in the diagonal series reaches the ground – so that there is now a short period during which only two feet are on the ground, instead of three. We can see this by watching the two left feet of a horse (Fig. 31); the forefoot is lifted just before the

77

hindfoot has reached the ground, and sometimes the hindfoot is put down actually over the footprint of the forefoot. Similarly, the left hindfoot of the horse is lifted before the right forefoot has reached the ground. During these short periods the body either rolls towards one side or tips diagonally forwards, as the case

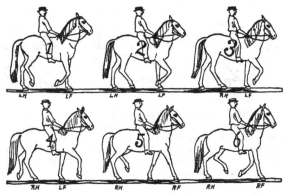

FIG. 31. *A walking horse balances its body alternately on two feet on the same side of the body and on a pair of diagonally opposed feet: in (1) both left feet are on the ground (Phase 1); in (2) the right hindfoot is coming down towards the ground, and in (3) the left hindfoot is about to lift, so that in (4) the body is supported by the diagonally-opposed left fore- and right hindlimbs (Phase 2). In (5) the right forefoot is coming down and the left forefoot about to lift, so that in (6) the body is balanced on the two right feet (Phase 3)*

may be, but these turning movements cannot go very far before the missing foot has come down and set matters right again. If the process of premature lifting of the feet is carried still further, a stage will be reached when there are never more than two feet on the ground at one time; the complete limb movement may then be divided into two equal periods, during

78

each of which the body moves forward a span. The distance between successive footprints is then twice the span; so that if the frequency of limb movements is still 1 per sec., and the span 3 in., the speed of progression is increased to about 600 yd per hr.

When an animal is using two limbs simultaneously

FIG. 32. *A trotting horse balances itself on diagonally-opposed feet. In the two top series of pictures (1–6) the left hind- and right forefeet are on the ground. In the last two pictures (8, 9), the right hind- and left forefeet are in action, whilst the other two limbs are swinging forward in the air*

in this way, the gait is completely unstable, in the sense that there is no moment in the whole movement when the legs are providing an adequate triangle of support for the animal's weight; the moving animal keeps its balance because each new pair of supporting legs comes into action soon enough to correct the instability caused by the previous pair. In other words, the 'two-pin' pattern of support can only be employed if the limb movements are quick enough.

79

When a horse or zebra is walking quickly, the whole cycle of its actions can be divided up into four stages of two-point suspension.

Phase 1 {Left hindfoot, Left forefoot;   Phase 3 {Right forefoot, Right hindfoot;

Phase 2 {Left forefoot, Right hindfoot;   Phase 4 {Right forefoot, Left hindfoot;

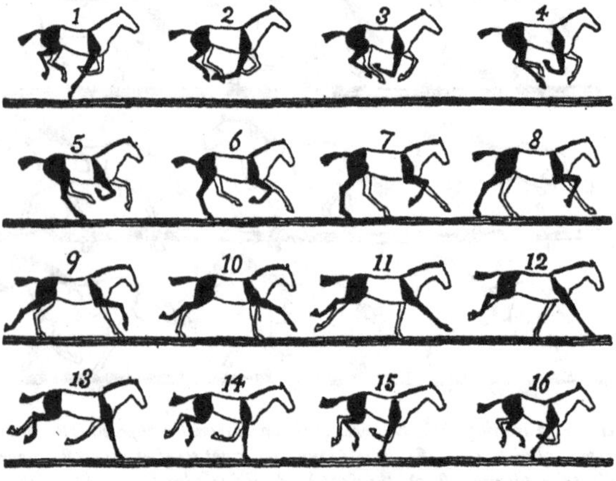

FIG. 33. *During a gallop there are never more than two feet on the groun simultaneously and during part of the time the animal is in the air. Note hou the limbs are flexed when swinging forward; the feet are moving backwara rapidly before they strike the ground*

As the limb movements quicken still further there is a marked tendency to shorten the period of unilateral support (RF, RH) and lengthen the period of diagonal support (RF, LH). This is particularly clear in the gait of such animals as newts and lizards, but it is also characteristic of a trotting horse (Fig. 32). A few animals accentuate the bilateral phases at the expense of the diagonal phases, and this kind of action is usually spoken of as the 'rack'.

As an animal changes from a three-point to a two-point suspension, the period of time during which each limb is on the ground decreases, partly because the time taken for a *complete* cycle of limb-movement is decreased, and partly because each limb is on the ground for a smaller fraction of the time required for one complete cycle. If a limb is on the ground for a shorter period than it is off the ground, there must be periods in a two-point gait when the whole animal is off the ground – in other words, it must bound or jump forward between each successive impact of the limbs with the ground.

If the process of using fewer limbs simultaneously were carried to its logical conclusion, there would be a type of motion in which only one limb was on the ground at a time, and between these periods there could be periods in which no foot is on the ground. This state of affairs is reached by a galloping horse (Fig. 33), when the distance between successive imprints of the same foot may be as much as 25 ft. It should be noted, however, that the sequence in which the limbs are moved is no longer the same as in a walk; a galloping horse lifts its feet either in the order LF, RF, LH, RH or RF, LF, RH, LH.

A galloping horse marks, perhaps, the end of one of Nature's great experiments, for here the power by which the horse is driven forwards comes almost entirely from the muscles that move the limbs; and relatively little from the muscles of the back, as in the case

81

of a fish. Between these two extremes there exist a few intermediate types. When a newt is running, only part of its propulsive energy comes from the pairs of diagonal limbs; part is still derived from the muscles of the back: the body undulates very much like that of a fish. Another obvious example of 'lumbar' propulsion is seen in a dog: a greyhound increases the

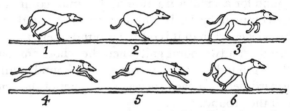

FIG. 34. *By arching and extending its back a galloping dog greatly increases the power and length of its stride*

span of its hindlimb by arching its back (Fig. 34), and the speed and power of these limbs is greatly increased by the contraction of the back muscles.

For speed and endurance over some distance – flat-racing qualities, you might say – there can be little doubt that animals such as antelopes and horses are the best performers – for they alone can keep up a speed of 40 or 50 m.p.h. for quite a long time. For shorter spells, however, the cat tribe (notably the cheetah) comes very close. Let us look at the limbs of these fast runners and see how they differ from those of slower animals, such as bears or monkeys.

The legs of a bear are, in comparison, heavy structures, with powerful muscles operating about each joint. Each leg ends in a foot, in structure and movement not greatly different from our own. Each foot has a large 'plantar', or bearing-surface. Heel and toe are both in contact with the ground during the first part of the limb's backward movement; and during the second half of this movement the heel is raised and the body swings forward on a fulcrum provided by the toes alone. This transference of weight to the toes occurs when the hip or shoulder has moved forward in front of the foot; and the raising of the heel not only reduces the strain falling on the muscles of the ankle joint but also has the effect of increasing the length of the animal's stride. It is important to notice that the moment of raising the heels comes when the feet lie behind the hip or shoulder joint.

Contrasted with a bear's leg, the leg of a horse shows three main differences (Fig. 35). First, the five toes in each limb of a bear help to provide a broad bearing-surface; but the limb of a horse consists of one toe only, and this ends in a small and rigid hoof. Secondly, neither the heel ('hock') nor the wrist ('knee') of a horse ever rests on the ground; these joints are carried well away from the ground because of the length of the bones uniting the 'finger' to the rest of the limb. Thirdly, the limbs of the horse are longer than those of a bear and the muscles that work it are concentrated at its upper ends, the lower joints being operated by long tendons. Unlike a bear or a man, a horse supports itself on the tip of the middle toe of each foot – all the other digits have disappeared. This type of foot is known as *digitigrade*.

Between the plantigrade limb of a bear and the

FIG. 35. *The limbs of a bear are not very different from our own; its feet have relatively large surfaces and the ankle joint is close to the ground when the limb is put down. The limbs of a horse are different, for the ankle ('hock') and wrist ('knee') joints are always held well away from the ground and the animal stands and walks on its 'toes'*

digitigrade limb of a horse a great variety of inter-mediate forms may be found. The hindfeet of a dog, a cat, or a rabbit are examples. These animals can all squat on their heels, but when they run, the heels are always raised as a horse's are, though the weight of the body is carried on the under surface of the finger bones and not on the tip of the one bone as in a horse. To what degree does the form of the horse's limbs help the animal to be able to run at high speed over hard, level, and smooth ground?

We know that no animal can apply a forward force to its body greater than the backward force which, at the same time, it applies to the ground with its feet. With a horse, the backward force that can be exerted by a limb is limited by the amount of friction between the hoof and the ground. On perfectly smooth ice, a horse could not go forward at all, for the feet would 'skid' as the back wheels of a car do. The amount of foot friction depends on the roughness of the ground and the weight carried by the leg. I am sure you have noticed that when a horse is startled it gives a sudden powerful backward push with its hindfeet against the ground, and the forelegs often rise from the ground altogether, as the animal bounds forward. Or again, the forefeet of a horse will often slip when the animal is trying to get going with a heavy load. This upward movement of the front of the body, and the slipping of the forefeet during intense effort, are a result of the backward thrust of the hindlimbs. The thrust acts at ground level, but the backward drag of the load and the inertia of the heavy body acts at the level of the shoulders and hips. The two forces together, therefore, form a 'turning couple', tending to lift the front end of the body. This lifting action at once reduces the weight carried by the forefeet, and correspondingly increases the weight being carried by the hindfeet. If a horse is pulling against a very heavy load, the amount of weight being carried by the forefeet is thus decreased, and therefore the amount of friction between forehoofs and the ground is not enough to allow the muscles of the forelegs to exert their full effort without slipping. In the case of the startled horse bounding forward, the upward-turning effect of the hindlimb is greater than can be compensated by a readjustment of weight from the fore- to hindfeet and, as a result, the front end of the body tilts upward and the front legs leave the ground. We now see why a horse should normally keep more weight on his forefeet than on his hindfeet, for in this way he can use all his four feet to their maximum forward moving strength.

When an antelope or a racehorse starts from rest, the whole of its muscular effort is being used to overcome the inertia of its body; the greater the backward thrust of the feet against the ground, the quicker does the body gather speed. But when this speed reaches about 45 m.p.h. no further increase occurs, although the animal is moving its limbs extremely rapidly. Why should this be? What limits the speed of such animals? Obviously wind resistance plays a part, and as wind resistance increases with increasing body-speed, a time must come when the backward drag of the wind equals the forward drive of the feet. But there is an even more important factor. If a foot is placed on the ground when its backward speed relative to the hip is less than the forward speed of the whole body relative to the ground, the foot must then act as a

86

brake, and the body drives the limb instead of the limb driving the body. The conditions for active propulsion are essentially the same as for the back wheels of a car; if the wheels are to drive the car, the power applied by the engine must be such that the wheels would go even faster if they were lifted off the ground. If, when the wheels are removed from the ground, they spin slower, then they must have been acting as a brake – in other words, the body was providing energy for the engine instead of the engine for the body. Just as the maximum speed of a car is determined by the frequency of revolution of its wheels, so the maximum speed of an animal depends on how frequently it can oscillate its limbs as pendulums.

Many fast-running animals do not in fact place a foot on the ground, until its backward speed relative to the hip has almost reached its maximum value at the midpoint of the swing: the foot comes down almost vertically under the hip or shoulder. At high speeds, the amount of energy required to accelerate the limbs about the hip and shoulder joints is probably far higher than that required to overcome air resistance. Some of these points are illustrated in Fig. 33, which shows various stages in the movement of the limbs of a galloping horse; you can see that, although the limbs project well ahead at the end of their forward swing, the hoof does not touch the ground until the backward swing of the leg is well under way. At this stage the hoof is travelling backwards relative to the body at a speed equal or greater than that at which the horse's body is travelling forwards. The effect of impact is to check the backward speed of the hoof (relative to the shoulder or hip) and not to increase it. You will also notice in Fig. 33 how the limb is flexed as soon as it is lifted from the ground. This 'high stepping' action is very important at high speeds, for it materially reduces the amount of energy required to swing the limb forward; in technical language, it reduces the 'moment of inertia' of the limb.

The same high-stepping principle is a feature of good human running (Plate 5). A good runner flexes his knee sharply before he swings his leg forward, and he does not put it to the ground until its backward speed relative to the hip is at least equal to that at which his body is travelling forward.

One other mechanical feature of human running or walking may be noticed here: the action of the arms. The natural actions of a man's arms and legs are essentially the same as those of a horse whose diagonal limbs are working together; in man the right arm swings forward with the left foot and the left arm with the right foot. In this way the tendency of a leg to make the body yaw to the opposite side is compensated: the more active the action of the leg, the greater should be the amplitude of swing of the arm.

We must not, of course, suppose that the speed which an animal or man can maintain depends on mechanical factors alone. Except in a short sprint it depends much more on the rate at which the heart and lungs can supply oxygen and food to the muscles, and the rate at which these muscles can thereby re-generate their supplies of energy.

88

Apart from this important factor, animal speed depends largely on the power of the muscle to operate the limbs as pendulums. On the other hand, the rate at which an animal can get going depends on the rate at which the muscles can be used for overcoming the inertia of the body. A rhinoceros can move as fast, or faster than a man, but probably takes longer to get under way.

If (as I hope you will) you observe the movements of animals for yourself, you must not be surprised or disappointed if you find exceptions to the rules which I have tried to describe. The basic principles will, I think, help you, but you must be prepared to find that some animals, and even individual animals, have their own peculiar gaits. You will find that your own dog will vary the movements of his legs quite considerably; and if you go to the Zoo, you might find it interesting to decide for yourself, whether or not, a camel's or a giraffe's legs move in the same order as those of a bear.

4

Jumping and Creeping

JUMPING

In athletic sports we judge a man's jump either by measuring the height (in feet and inches) to which he raises his body above level ground, or by measuring how far forward he travels in his jump before landing on the earth again. In both cases the body rises from

Animal	Standing high jump		Long jump	
	Jump (in feet and inches)	Jump (in units of body-length)	Jump (in feet and inches)	Jump (in units of body-length)
Kangaroo	8 ft	1·5	26 ft	5
Frog	9 in.	3	3 ft	12
Grasshopper	8 in.	18	–	–
Flea	6 in.	100	12 in.	200

the ground because the legs exert an upward thrust against the body greater than the downward pull of gravity. The difference between these two forces (the upward thrust and the downward pull) is available for accelerating the body upwards against its own inertia.

Almost all animals with legs can jump more or less, but if we wanted to choose a group of really good animal jumpers we should probably include a kangaroo, a frog, a grasshopper, and a flea. The table shows at once that a flea's long jump is 200 times its own length, but that a kangaroo's long jump is only 5 times its own length. We should assume, of course, that they would all be allowed to take off from firm ground, and that in comparing them as jumpers, allowance would be made for their very different sizes. In an 'open' event a kangaroo would be an easy winner; but if we measured the height or length of a jump not in feet and inches but in terms of the length of the jumper's own body, a flea would be found to be a far better jumper for its size than a kangaroo.

Let us try to find out why the four animals we have chosen are so much better as jumpers than most other animals of comparable sizes. One thing is obvious at once. All good high jumpers have relatively long and slim-built legs. Just before a standing take-off all the joints of the legs are tightly folded up under the body, but as soon as the jump begins the joints extend to the full by the shortening of the muscles at the upper end of the limbs. A good example is shown in Plate 7, which shows a locust 'hopper' jumping vertically upwards.

As soon as the animal leaves the ground it has to rely on its own kinetic energy to overcome the downward pull of gravity, and as this kinetic energy depends on the mass of the body and its speed, the height of a jump is determined by the speed at which the body leaves the ground. Newton's Law tells us that if a body of mass m leaves the ground at a velocity V, its kinetic energy is $mV^2/2$; if the whole of this energy is used up in raising the body against gravity, the body's upward speed will be reduced by a constant amount (g) during every second of its upward flight and the body will cease to rise after V/g seconds, because at that moment all its kinetic energy has been used up in lifting the weight of the body through distance (h). The height of the jump is related to the take-off speed by the expression $h = V^2/2g$ or, if we are expressing distance in feet, weight in pounds, and time in seconds, $h = V^2/64$: thus to jump 1 ft high, the take-off speed must be 8 ft per sec.; and to jump 9 ft, the take-off speed must be 24 ft per sec. It will be noticed that the take-off speed does not depend on the weight or size of the animal. To jump 1 ft into the air the take-off speed of a flea is the same as that of a man. In practice, the take-off speeds have to be rather higher than the calculated value, because some of the kinetic energy of the body is used for overcoming the resistance of the air; for the present, however, we can focus our attention upon what the legs of an animal must do to propel the body off the ground at a given speed.

From the moment at which the muscles of the legs begin to shorten, until the moment when the animal leaves the ground, the trunk of the body, starting from rest, has reached a velocity V. This change of speed has come about whilst the trunk has risen a distance (s) equal to the extent to which the tops of the limbs have moved upwards. To do this, the trunk has to be subjected to an acceleration (a) related to the take-off velocity by the expression $a = V^2/2s$. In order to provide this acceleration the limbs of the animal have to exert (during the time they are extending) an average force (F) against the body and against the ground such that $F = ma$, where m is the mass of the body. As $a = V^2/2s$, $F = mV^2/2s$ and this force is over and above that required to support the weight of the body. The total thrust of the legs during the take-off is thus $(mV^2/2s + mg)$. In other words, a given take-off speed (and therefore a given height of jump) requires a smaller thrust from the limbs when (i) the legs are able to extend to a greater length (s), and (ii) when the mass of the body (and consequently its weight) is small. Notice that during the take-off period the body and the upper ends of the limbs are travelling upwards much faster than the lower parts of the limbs; in fact, until the body leaves the ground, the feet are not moving upwards at all. At the instant of leaving the ground entirely, some of the kinetic energy of the trunk and the upper joints of the legs has to be used for lifting the feet, therefore the speed at which the whole body would leave the ground would be con-siderably reduced if the feet and lower limb-joints were heavily built: the animal would, as it were, be trying to jump in heavy hob-nailed boots.

If we apply these principles to the locust shown in Plate 7, we find that the hindlimbs have extended through a distance of about 1 in. in $\frac{1}{40}$ sec., and con-sequently an acceleration of nearly 270 ft per sec. per sec. must have been applied to the body; this means that the limbs applied to the body (and to the ground) a force equal to eight or nine times the weight of the body. The take-off speed of the hopper was about 7 ft

93

per sec., so that the height of the jump ought to be 8 in.; this was what it was, in fact. An energetic hopper could, no doubt, do a good deal better than this.

Most animals make use of their jumping powers to get out of danger quickly, and for this a long jump is likely to be more effective than a high jump. An ordinary frog will serve as a good example of a standing long jumper. The photographs in Plate 6 show a frog which propelled itself forwards as well as upwards by extending its back legs to their fullest extent in a direction slanting upwards by an angle of about 45°. After leaving the ground, the frog followed a curved track and finally landed on its front feet about 20 in. away from the spot where it left the ground. At the moment of the take-off it was travelling upwards at about 4 ft per sec., and forwards at the same speed. Now, the upward speed is steadily decreased by the pull of gravity; the forward speed remains unchanged during the whole jump. As the initial upward speed was 4 ft per sec., against the downward acceleration of gravity of 32 ft per sec. per sec., the frog's upward flight ceased after $\frac{1}{8}$ sec., and at the end of this time its centre of gravity had risen 3 in. above its take-off position and about $7\frac{1}{2}$ in. above the ground.

$$
\begin{array}{ll}
gt = V & 2gh = V^2 \\
32t = 4 & h = (4)^2/64 \\
t = \tfrac{1}{8} \text{ sec.} & = \tfrac{1}{4} \text{ ft} = 3 \text{ in.}
\end{array}
$$

Having reached its highest point, the animal began to fall under the force of gravity and reached the ground again in about $\frac{1}{8}$ sec. From the moment of take-off at point T, to the moment of landing, the frog was in the air for a total time of $\frac{1}{8} + \frac{1}{8}$ sec., and as it was travelling forward at 4 ft per sec., the length of the jump from the take-off at T to the landing point was about 16 in., or 20 in. from the point at which the feet left the ground. This example shows that the length of a standing long jump depends on two factors: (i) the time that the animal can keep itself in the air (in other words, the height to which it jumps) and (ii) the speed at which it is travelling forwards at the moment of take-off. In Plate 6 the frog took off at an angle of 45° and its upward speed was therefore the same as its forward speed; half the energy of the leg muscles was used for driving the body upwards and half was used for driving the body forwards. This particular division of effort results in the longest jump; the jump will be shorter if either more, or less, than one-half of the total effort is used for accelerating the body forwards instead of raising it in the air. It would be interesting to know whether all frogs always jump as skilfully as the one in the photographs, or whether this particular frog happened to be more lucky or more expert than some of its brothers and sisters.

Before leaving the frog in Plate 6, let us work out how high it would have risen if it had put the whole of its effort into a *vertical* jump. It would have had twice the amount of energy to propel itself against gravity, and its centre of gravity would therefore have risen twice as high above its take-off position; a 6 in. rise of the centre of gravity would represent an effective jump of more than 10 in. above the ground. To make

this jump (or the one shown in Plate 7), the total force
exerted by the frog against the ground must be about
three times the weight of its body. Had the frog, like
the hopper (shown in Plate 7) been able to exert a
force equal to *nine* times its own weight, it would have
risen 18 in. above its take-off point. But until a frog
has shown that it can do this, I am inclined to suspect
that it can't – that such a jump is a good deal beyond
its powers; so from this point of view the hopper is a
better high jumper than a frog.

If we measure the height or length of a jump in
units of the animal's length, the achievements of a
frog, and even of a hopper, are small when compared
with those of a flea. It is, of course, very hard to observe and measure the movements of fleas, but the
most reliable evidence suggests that the more expert
among them can reach a height of about 8 in. in a
vertical jump, and that the range for a good long
jump is about 12 in. If the length of the flea's body and
the range of extension of its hindlimb were both about
$\frac{1}{20}$ in., its take-off speed for a vertical jump of 8 in.
would have to be about $6\frac{1}{2}$ ft per sec.; and to produce
this speed, the flea's hindlegs must exert an average
force against the ground equal to 160 times the weight
of its body. Such an effort is obviously much greater
than that of a locust, and far beyond that of a frog.
The relatively poor performances of the larger animals
become still more obvious when we look at a kangaroo.

So far as we know, a large kangaroo cannot make a
standing high jump of much more than 8 ft, although
the possible range of extension of its hindlegs is prob-

ably not much less than 4 ft. If these estimates are correct, the force available for accelerating the body is twice the animal's weight. If a 5 ft kangaroo could jump as many times the length of its own body as a flea can, it ought to be able to raise itself to a height of 800 ft, or, as Professor A. V. Hill has put it, 'to jump over St Paul's'. To perform this feat, the kangaroo's take-off speed would have to be 224 ft per sec., or about 150 m.p.h., compared with the 4½ m.p.h. of a flea when jumping 8 in. Similarly, the force that would have to be exerted by the feet of a kangaroo weighing, perhaps, 100 lb. would be about two-thirds of a ton. A force as big as this would almost certainly break the hindlegs; and even if the take-off caused it no damage, the jumping kangaroo would certainly be dashed to death when landing again at 150 m.p.h. These things do not happen, because the muscles of a kangaroo's legs, powerful as they are, are not powerful enough to raise the body at the necessary speed against its own weight.

'Then what happens in the case of a flea?' you may ask. To answer this, it is useful to consider what effect body-size has upon the rate at which the muscles of the leg must generate energy in order to lift the body for a height equal to its own length. The energy required for the jump (weight × height of jump) must be proportional to the fourth power of the length of the body, because the weight is proportional to the cube of its length. On the other hand, the weight of the leg muscles varies only with the *cube* of the body-length, and consequently to effect a jump of the same

G   97

height relative to body-length, each gram of muscle would have to contribute an amount of energy proportional to the length of the body. As a kangaroo is many hundred times larger than a flea a kangaroo could only jump over St Paul's if each gram of its muscles provided, during the take-off, many hundreds of times the amount of energy produced by a gram of flea's muscle when this animal jumps 8 in. On the other hand, if the muscles of a kangaroo could only contribute as much energy as that of a corresponding weight of flea muscle, a kangaroo could only jump 8 in.! This is clearly not the case, and we have to conclude that, weight for weight, a kangaroo's muscles are more powerful than those of a flea.

Professor Hill has shown, however, that the output of power from a muscle depends on the speed at which the muscle can shorten when it is not lifting a load, as well as on the load which it has to lift. Some muscles shorten quickly when not loaded; and others more slowly; but, in general, when not lifting a load, small muscles (and corresponding muscles in small animals), can shorten much more quickly than larger muscles. If any muscle is to give its best possible amount of useful energy, it must shorten under load at about one-third of the rate at which it would shorten under no load, and this condition is satisfied when the load is about one-third of what would just be necessary to prevent the muscle shortening at all. A kangaroo could – to use Professor Hill's own words – jump over St Paul's 'if his muscles were intrinsically as quick as the grasshopper's and 100 times as strong as they are, and if his structure would stand the accelerations involved in taking off and landing! These conditions are all impossible: an animal designed for high jumping with a reasonable factor of safety should in fact be able to jump the same height regardless of size, the larger animal having the smaller acceleration and the longer take off.'

Having looked at a variety of jumping animals, we may find it interesting to watch and consider the performance of a human being. During a standing jump the movements of a man's legs are not greatly different from those of animals, but we mostly use a running jump, where the conditions are somewhat different. After a short and not very fast 'run-up' an expert high jumper can clear a bar 6 ft from the ground, and during this jump he raises the centre of gravity of his body about 3 ft. According to cinematograph pictures the take-off has three stages:

(i) One leg (the left in Fig. 36 (5)) is fully extended forwards, with the heel on the ground, and the other (the right) is fully retracted backwards, together with the left arm.

(ii) The left leg, with knee slightly bent, pivots forward about the foot while the right leg and both arms are rapidly swung forwards and upwards (7–9).

(iii) With the arms raised, and the right leg nearly horizontal, the left leg straightens at knee and hip, and the body is lifted off the ground.

Altogether the whole cycle is quite complex, but certain features seem fairly clear: (i) The take-off limb

99

(the left in Fig. 36) first acts as a fulcrum on which the whole body is swung upwards and forwards, partly by its own hip muscles and partly by the kinetic energy

FIG. 36. *A good high jumper uses his take-off leg as a lever to swing his body upwards and forwards (5–9); swings his other leg and his arms upwards before taking off (7–9); and extends the knee of his take-off leg whilst the other leg and the arms are still travelling upwards (9)*

acquired during the run-up. (ii) During the second phase, the take-off limb presses downwards against the ground with a force that is not only enough to support the weight of the body but is also enough to enable the other leg and both arms to be accelerated upwards. (iii) The final straightening of the take-off knee does not begin until the centre of gravity of the whole body has been partly raised already by the lifting of the other leg and the arms.

The main principles of the jump seem to be to move the arms and one leg upwards whilst balanced on one

FIG. 37. *In a running long jump some of the kinetic energy gained during the initial run is probably used for propelling the body upwards. The man's trajectory is a flattened version of that which follows the standing jump of the frog*

foot, and then to apply a very powerful upward thrust from the take-off knee and hip. The whole cycle obviously requires very careful timing to ensure that the final upward thrust from the knee occurs when the other leg and both arms are travelling upwards at their maximum speed.

The record of a standing long jump is approximately 12 ft; I have no suitable photographs to reproduce here, but we may assume that the technique does not differ in principle from that of a frog. In a running long jump, however, the range reached by good athletes is about twice that of a standing long jump, that is, 25 or 26 ft. A running jump is really a high jump taken whilst sprinting as fast as possible. As seen in Fig. 37 the trajectory is a flattened version of Plate 6. If a man sprinting at 30 ft per sec. leaps into the air without checking his forward speed, the range of his horizontal jump will depend on how high he raises his body off the ground; for on this depends the period of time of his flight. If his centre of gravity rises 3 ft, the total time of his flight in the air would be rather more than 1 sec., and the length of his jump would be rather more than 30 ft. This is more than has yet been reached even by very expert athletes.

The drawings in Fig. 37 (although they are admittedly not entirely reliable) suggest that the man's forward speed is checked, slightly but perceptibly, at the take-off; it seems likely that some of the forward kinetic energy lost in this check of speed is used for raising the body by operating the take-off limb as a vaulting pole.

So far we have assumed that an animal – or a man – has been jumping from firm ground. But now let us go back for a moment to a salmon jumping out of water (Plate 3). The salmon spends most of its life in the sea, but returns to fresh water at breeding time. Every spring and autumn the adult fish swim into rivers and begin to make their way upstream, seeking to lay their eggs in the gravelly bed of the streams. As long as the river-bed is not too steep the fish swim through the water from one pool to another at speeds which are probably a good deal less than 10 m.p.h.; but when they come to a waterfall their progress is checked and they are confronted with the problem of getting over the falls. Near such falls, you can often see salmon jumping out of the water, and some can be seen to get over the fall. Others, however, fail to reach the top of the fall, and drop back into the pool. The height of these jumps varies; vertical leaps of 6 ft are fairly common, and as much as 10 ft has been recorded. Now, to reach a height of 6 ft, the fish must leave the water with an upward vertical velocity of about 20 ft per sec. (14 m.p.h.), and to jump 9 ft its upward take-off speed must be 24 ft per sec. (17 m.p.h.). To leap over the fall, the fish must jump forwards as well as upwards, and its take-off speed would thus have to be even greater. Perhaps a fish in good condition might attain such take-off speeds; but before we decide that such speeds are essential if the fish is to surmount the fall, we ought to give a little more thought to the water from which the fish leaps out. If this water were at rest, not flowing over the bed of the stream, the speed of the fish over the earth (the river-bed) would be the same as its speed through the water; thus a fish swimming upwards along a line inclined at 45° would come out from the surface and begin to travel through the air at the same angle. If it took off along such a line at 27 ft per sec., it would reach a height of 6 ft by the time ($\frac{3}{5}$ sec.) it had travelled 12 ft forwards through the air, at a speed of about 20 ft per sec. But the water at the foot of the pool is not likely to be still; it is probably flowing quite fast downstream; if it were travelling downstream at 10 ft per sec., a fish whose forward speed through the water is 20 ft per sec. will only be travelling at 10 ft per sec. in the direction of the fall – which means that it would have to take off at 6 ft from the fall, and not 12 ft. Unless the fish knows, at the moment of take-off, his distance from the fall and the downstream speed of the water, how could he judge the necessary speed and angular direction of his take-off? Remembering this, and also the high take-off speeds that would be necessary, I am driven to doubt whether many of the fish we see jumping are likely to succeed. It seems more likely that most fish surmount the fall without coming out of the water. Suppose the water is coming down the fall at 15 ft per sec. and that the fish enters this water at a speed of 20 ft per sec., he can then pass over a 6 ft fall at 5 ft per sec. if only he can keep up this speed for rather more than 1 sec.; moreover, while he stays in the water he will not have to lift his body against gravity, and he is able to make use of his muscles during the whole of his ascent.

CREEPING

The pentadactyl limb in one or other of its many forms has made it possible for vertebrate animals to move about over the earth on almost all kinds of surfaces; but thick undergrowth impedes such limbs and loose dry sand does not give a good foothold. For animals to move in such conditions as these, limbs would be little use and an older principle – propulsion by the backbone and its muscles – is employed. Snakes will serve as a good example of backbone propulsion.

When a grass-snake is placed on a very smooth, level surface it actively bends its body from side to side, but does not manage to make much progress. On grassy ground or on a gravel path it glides forwards gracefully, at a seemingly high speed, though this, in fact, is seldom more than 4 m.p.h. As a snake moves forward, every part of its body moves forward at the same time and follows faithfully along the same sinuous path as that of the region next ahead of itself. The forward motion depends upon two factors: (i) the sinuous or serpentine form of the body; and (ii) the presence of projections from the ground against which

FIG. 38. *When a piece of rubber tubing is drawn through a curved glass tube, the rubber is stretched on the outside of the bend and compressed on the inner side. Similarly the muscles of a snake lengthen on the outside of the bend and shorten on the inner side*

the snake can brace the sides of its body (see Plate 9).

What is essential to snake movement can readily be watched by sliding a length of marked rubber tubing through a glass tube bent into sinuous curves (Fig. 38). Equidistant rings are drawn on a length of the rubber tubing. If we watch the rubber moving along the curves of the glass tube, it is easy to see that the lines of these rings come closer together when the rubber tubing approaches the 'inner' surface of a bend in the sinuous glass tube, and further apart as it approaches

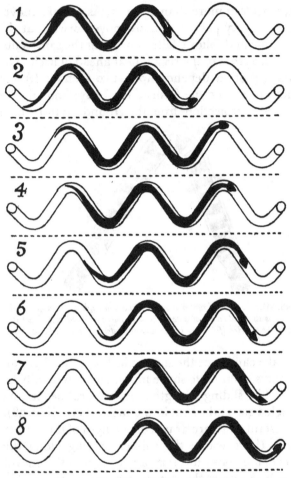

FIG. 39. *A grass-snake readily glides through a sinuous glass tube. It cannot glide round the arc of a circle or along a straight line*

the 'outer' surface of a bend. If we watch two adjacent lines as they pass along the tube, we see that each region of the rubber tubing is alternately bending first towards its left side and then towards its right side, and is continuously changing its shape into the shape previously displayed by the part of the tubing immediately ahead of it. Precisely similar changes of form are undergone by each successive region of a snake's body when it glides by its own forces through the same glass tube (Fig. 39); but with the snake, the changes in shape of the body are due to active changes in the length of the muscles which run down the body on each side of the animal's backbone. When a region of the body moves along the glass tube from an outer crest to an inner crest, the muscles on one side of the body shorten, whilst those on the other side are stretched. Muscles, we know, can only provide energy for propulsion by shortening in length; it follows that the serpentine glide can only occur when each part of the body moves along a path whose curvature is constantly changing. This very important fact becomes clear when we draw a piece of ring-marked rubber tubing through a glass tube bent into the arc of a circle, or through a glass tube that is perfectly straight; under these conditions there is no change in shape as the tubing is drawn along. A snake cannot *glide* round the arc of a circle or along a perfectly straight line.

When the snake is enclosed in a glass tube, the bending action of the body-muscles is opposed by the rigid walls of the tube. This all-important function of the glass tube can be clearly shown by substituting for the living snake two rigid rods hinged together by an elastic spring (Plate 8). When not opposed in any way, tension in the spring tends only to flex the ends of the two rods about their hinge, but if this movement is opposed by a rigid sinuous track (comparable with the original glass tube), the two rods glide automatically forwards. No such movement takes place if the track is circular or straight.

It should be noted that the forces exerted by the snake against the outside world (and those of the rods against the track in Plate 8) act at right angles to the body of the snake. During a steady glide the resultant of all these 'normal' or perpendicular forces provides a forward thrust equal but opposite to the frictional or tangential forces opposing the snake's motion. So long as the projections from the ground are rigid, the smoother they are the easier can the snake progress.

Although a grass-snake cannot glide along a straight line, it does not follow that it cannot escape from a straight rigid tube. It escapes readily enough provided the tube is not too narrow. When a snake is confronted with the problem of moving along a straight rigid channel it uses an entirely different method of moving along, as shown in Fig. 40. It makes its body into a 'concertina'; one end is thrown into folds, the outer ends of which are firmly pressed against the walls of the tube, and then the other end of the body can be drawn towards or pushes away from these fixed points. You will see, of course, that a snake could not move in this way in a *perfectly* smooth tube because the animal relies on friction to hold the folded

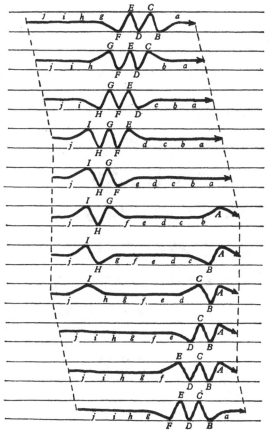

FIG. 40. *When enclosed within a straight or circular tube the animal progresses by 'concertina' movements. Capital letters indicate regions of the body which are at rest with the 'bellows' of the concertina tightly pressed against the walls of the tube and forming fixed points, against which the animal is able to exert a backward push sufficient to push or pull the rest of the body forward. The regions of the body which are in forward motion are shown by small letters*

part of its body firmly against the walls.

A grass-snake cannot move when its body is forced to form a straight line or the arc of a circle. Some snakes however – in particular, pythons and boa constrictors – can perform this feat. These snakes have a

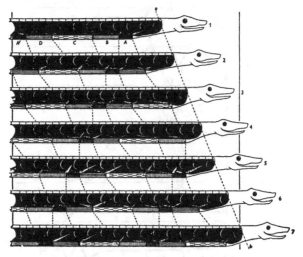

FIG. 41. *Some snakes (e.g. boas, pythons) can move along with their bodies forming a straight line. These animals possess a powerful band of muscles running along their ventral surfaces. Waves of shortening pass along this band from head to tail. These muscles, together with the ventral scales, enable the under surface of the body to operate as an enormous earthworm carrying the rest of the body on its back*

powerful band of muscle, attached by its under surface to the ventral scales of the body (the scales that touch the ground) and by its upper surface to the vertebral column. As the great snake moves along, waves of shortening pass along this band of ventral musculature from the head-end to the tail-end of the body (Fig. 41).

When each region shortens in length, it tends to pull the front end of the body backwards and the hind end forwards; the front end does not move backwards because the scales on the belly of the animal engage firmly with the ground; consequently, the region of the body behind the contracting muscles is drawn forwards. The under surface of the body thus behaves like an enormous earthworm, each region moving forwards as the underlying muscles shorten or lengthen, and remaining fixed to the ground when the muscles are fully shortened; the upper part of the snake's body is rigid, so that the whole animal may be likened to an earthworm bearing a rigid rod on its back. Straight-line movement of this kind, however, is relatively slow; when the boa or python means business it moves forward by means of its lateral muscles in a typical serpentine glide.

The three types of motion so far considered – the serpentine glide, the concertina motion, and the rectilinear glide – have one feature in common: the snake moves onwards along a line joining its head and tail. In some snakes, however, notably the 'side-winders' and puff-adders, the animal as a whole moves to one side of the head-and-tail axis. Snakes of this kind use the device of a caterpillar track, and their motion is illustrated in Fig. 42. Their muscular contractions are very similar to those of a serpentine glide, but each segment of the body remains at rest on the ground when its muscles are of equal length on the two sides of the body; a series of such segments form a fixed track, paying out segments at one end and laying down new ones at the other. The whole body thus moves sideways, leaving a series of parallel lines on the ground.

Snakes are not the only animals that can creep or glide. Many lizards glide under suitable conditions, and amongst animals of the lizard kind we find a wide variety in the size of the limbs; in some the limbs are well developed, in others (the skinks) the legs are quite small, and the slow-worm (*Anguis*) has no legs at all. Many of these gliding lizards are expert burrowers, and the muscular mechanism they employ is the same in essence as that of a snake passing through a sinuous tube.

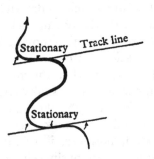

FIG. 42. *Other snakes (e.g. 'side winders') use the principle of a caterpillar tread. Two regions of the body remain stationary on the ground and form the 'tread'. Successive regions of the body are added to the front of the tread, whose hinder end is being lifted. The whole snake thus moves sideways, leaving a series of parallel track lines on the ground*

Who knows at what speed snakes can move? Some are said to travel 'faster than a horse', and no doubt a snake can move faster than a horse, through very thick jungle. But can a snake really get along as fast as a horse, on open ground? Such claims must be treated with suspicion until we have something better than hearsay to go upon. It is not easy to make any sound guess at its speed unless a moving snake can be watched for some time; and to keep an eye on a snake as it travels through thick grass is not an easy matter. The great naturalist W. H. Hudson gives the following description of a fast-moving snake. 'The creature, as one looks down on it, changes its appearance from a narrow body moving in a sinuous line to a broad straight band, the outward and inward curves of the body appearing as curved lines on its surface, and the spots and blotches of colour forming the pattern as shorter lines. The shallow pebbly current shows a similar pattern on its swiftly moving surface, the ripples appearing as light and dark slanting lines that intersect, cross, and mingle with each other. ... Viewed from an elevation, all rivers winding through the lower levels, glistening amidst the greens and greys and browns of earth, suggest the serpent form and appear like endless serpents lying across the world.'

H

5

Flying Animals

IF a kangaroo, say, or a frog, or a flea, or any land-living animal, is to lift its body off the ground at all and stay in the air even for a very short time, we know that it has to exert an intense muscular effort to do so. Then how can a bird rise so easily from the ground and stay in the air in flight for hours at a time?

Flight depends on wings. A wing, we might say, is a limb whose movement through the air produces forces that can counteract the downward pull of gravity, and can also drive the body forwards through the air. It has long been known that wings can do these things, and from time to time adventurous people have tried to design mechanical wings capable of lifting a man and carrying him along through the air. All these experiments failed; but in the end, for good or ill, they led on to the invention of aeroplanes. Men tried for a long time, by watching the birds, to learn how a man could fly; to-day, quite the opposite, we are trying to understand the flight of birds by applying principles which have emerged during the design of aeroplanes. The movements of a bird's or an insect's wings are extremely complicated, and it is easier to feel our way into the very difficult problem of animal flight by drawing a distinction between two kinds of flight –

active *flapping* flight; and passive *gliding* flight.

We can start our inquiry, then, by comparing the motion of a soaring eagle with that of a 'glider' aeroplane; in both, the wings are used as fixed and rigid surfaces, and neither glider nor eagle uses an internal engine or source of power. From the very start of our study we must realize that all flight – whether active or gliding – depends on forces set up between the wing and the surrounding air. In a vacuum, an aeroplane or a bird would fall to the ground just as rapidly as a stone. We must also understand that the air only exerts a force against the wing when there is movement between them – either by the wing moving through the air or by the air moving past the wing.

In its simplest form a wing can be thought of as a flat plate sticking out sideways from the body of a bird or from the fuselage of an aeroplane. If this flat plate be held edgewise in a current of air, the force of the current will tend to drag the plate with it, downstream; and only an external force tending to pull it upstream can hold it in position. If, however, we slightly raise the edge that faces upstream (or lower the downstream edge) the plate will tend to rise bodily in the air, although, at the same time, the force tending to drag it along downstream increases. The moving air is thus exerting two forces on the inclined plate (Fig. 43): (i) a *lift* force tending to raise the plate; and (ii) a *drag* force pushing it back in the direction of the air stream. True gliding flight becomes possible when these two forces are so adjusted as to be exactly equal to the weight of the bird or aeroplane.

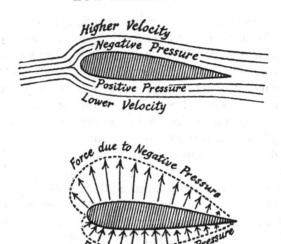

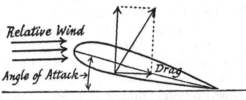

FIG. 43. *When a wing is held at a slight angle to an air current the air flows faster over the upper surface than the lower, thereby creating suction above the wing and pressure below the wing; the suction and pressure together cause the 'lift'. At the same time the moving air tends to 'drag' the wing backwards. The total effect is to lift the wing and drag it backwards*

Tests with various sizes and shapes of wings, in winds of various speeds, have shown that the lifting effect of the air depends on four main factors: the shape of the wing; its size; the angle at which it is inclined to the direction of the wind; and the speed at which the wind is travelling past it.

The mechanical efficiency of a wing depends largely on its ability to develop a large lifting force for a relatively small increase in backward drag; but, from a practical point of view, the ability to fly depends on the extent to which the exact balance can be maintained between the lift and drag forces of the air on the one hand, and the pull of gravity on the other. Stable, continuous flight can only happen when the wings are so arranged that any slight accidental disturbance of the air-flow is automatically compensated.

We can see these principles in action when a thin sheet of light wood or card is released in the air. A uniform sheet glides for a short time towards one side, then changes its direction of motion and falls to the ground like a falling leaf. But if a small lead weight is attached to the leading edge, the sheet glides smoothly downwards when the size of the weight is so adjusted as to bring the centre of gravity of the whole 'wing' to a point about twice as far from the trailing edge as it is from the leading edge. We can, with a little practice, get the same result – a smooth glide downwards – by shaping the card itself into the outline of a bird with outstretched wings and tail.

We have now reached a point at which we can say that an animal can fly, provided it keeps its wings moving through the air in such a way as to ensure that the lift and drag forces exerted by the air against the body all combine to form an upward force equal to the animal's weight and acting through the centre of gravity of its body. Now let us look at some of Nature's gliders.

Broadly speaking, gliding animals may be separated into two classes, those that develop the necessary motion through the air by their own muscular efforts beforehand, and those that keep themselves moving by falling under gravity. Flying fish (*Exocoetidae*) are good examples of the first kind. What is noticeable about flying fish is the enormous size of their pectoral fins. When a flying fish is swimming under the surface of the water it keeps these fins furled against its sides, but spreads them wide as it leaves the water to fly through the air. Before the 'take-off' the fish drives itself slantingly upwards to the surface by vigorously swimming with its tail; as it leaves the water its pectoral fins (and sometimes also the pelvic fins) are spread, and the fish finds itself airborne just above the surface of the sea (Plate 10).

According to most observers, flight seldom lasts for more than 1 or 2 sec., and the distance travelled in that time varies from 10–50 metres. From time to time the fish may regain flying speed by beating its tail against the water, although the rest of its body remains in the air. There is, however, some evidence to suggest that much longer flights may occur, lasting for 10–12 sec., and covering perhaps 400 m. without touching the water. All observers agree that the fish does not move its pectoral fins in flight, though they may vibrate passively as it speeds through the air. About these flights we have much to learn, and exact observation is difficult, but there seems little doubt that ordinary flights of 1 or 2 sec. and of 50 m. or less are maintained by the kinetic energy (the energy of speed) given to the body in swimming by the tail striking against the water before the fish is airborne. The lift and drag of the body at various speeds of airflow have been measured in a wind-tunnel, and although such observations are not yet altogether satisfactory, they suggest that flight could be sustained for about 6 sec., and over a distance of 200 ft, if the take-off speed were something like 30–35 m.p.h. This speed seems very high compared with that of other fishes, but it may be that when the body of the fish is in air and only the tail is in water these higher take-off speeds could be reached. Still, it is difficult to believe that the take-off speed could ever be great enough for the fish to fly for 10 or 12 sec., and for distances of 400 m. If such flights really occur, as they are said to do, we can only assume that the fish is able to draw on some source of energy other than that given to it (when it is in the water) by its tail. It may be that unusual air conditions near the surface of the waves make longer flights possible; but all we can say at present is that long-sustained flights offer an interesting and difficult problem.

For the time being, however, we can look upon the normal short flight of the flying fish as a typical example of 'velocity' gliding: the fish stores kinetic. energy in its body by swimming – by moving its tail in water; and expends this energy when it is airborne in overcoming the drag effect of the air, and gaining enough lift to overcome gravity. And after all, this is not so very different from the flight of a small bird, where periods of passive glide occur between periods of active wing-beats.

One member of the frog family is a glider; the only one, so far as is known. This frog (*Rhacophorus*) was discovered in Borneo about a hundred years ago, by a Chinese workman employed by the naturalist, Alfred Russell Wallace. The Chinese assured Wallace that 'it came down, in a slanting direction, from a high tree'. Wallace noted that the surface of the very large webbed feet was considerably larger than that of the body, and there can be little doubt that these frogs do launch themselves from trees to glide down to a point thirty or forty yards from where they started. The webbed feet and the under-surface of the body act together as a wing, and the energy required to drive the animal along in the air comes from gravity in a way which we shall consider later.

Among living reptiles, only one rather doubtful glider is known – the so-called flying dragon (*Draco volens*); this odd lizard has a flexible membrane down each side of its body, and it is said to glide, like the flying frog, by jumping off trees. It seems strange that there are no other living flying reptiles, when one remembers the great pterodactyls of prehistoric times. Pterodactyls (Fig. 44) (meaning wing fingered) had enormously enlarged fourth fingers from which membranes spread across to the body and hindlegs: their wings sometimes had a span of twenty feet. If you want to imagine what they looked like you cannot do better than read A. Conan Doyle's *Lost World*. Unlike birds, pterodactyls had poorly developed breast muscles; active flapping flight must have been impossible; how these gliding monsters launched themselves, either from trees or from the tops of steep cliffs, we do not know.

And now we come to the greatest of all flying ani-

FIG. 44. *The pterodactyl, showing the enormously enlarged fourth fingers, with membranes extending to the body and hindlegs to form wings*

mals, the birds. The mastery of birds over air is incomparably greater than that of any other group of animals, and we shall have to examine them in detail a little later. Here we need only note that a bird's front limbs have been completely specialized for flight; each wing forms a structure of peculiar beauty and complexity (Plate 11). Unlike that of any other flying animal, the wing surface in a bird is made up of feathers, all fitting together to form an efficient lifting surface and yet capable of being neatly furled when not in use. There are two chief kinds of gliding birds: the *low gliders*, such as shearwaters and albatrosses; and the *high soarers*, such as eagles and vultures.

Of mammals, the best known gliders are the bats. A bat's wing surface is probably not unlike that of a pterodactyl, but instead of spreading from the fourth finger only, it is spread between the body and all the fingers of the hand except the thumb. A bat's flight is not very different from a bird's flight; but it differs in this, that bats normally rest by *day* and fly by *night*. How they do this we shall see later on.

There are also a number of mammals capable of gliding from tree to tree: flying phalangers, flying squirrels; all of these have membranes of some kind running between their wrists, body, and back legs. When such an animal is walking, the membrane is slack, but when it takes to the air for a glide, the front- and hindlimbs are spreadeagled, stretching the membrane out into a taut and effective wing (Plate 12 and Fig. 45). All these animals live in trees; they climb a tree, and from there, launching themselves into the air, glide gracefully down, often rising towards the end of their flight and landing on a neighbouring tree-trunk or branch.

All the flying animals we have examined so far have solved the problem of gliding flight only to the extent of being able to stay in the air for a few seconds. How do some birds glide or soar as they do for many minutes or even hours at a time?

Let us start with the type of glide seen when a

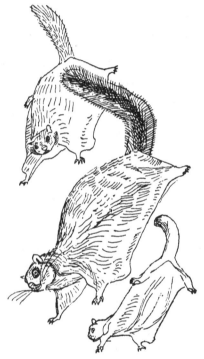

FIG. 45. *The flying squirrel* (Glaucomys), *showing its gliding flight with limbs outstretched*

pheasant or a gull is approaching the ground; the wings are stiffly stretched out and the bird is slowly losing height all the time. Its speed through the air is being maintained by the accelerating effect of gravity, and its motion can be likened to that of a sledge travelling down a slope. On a slope, the weight of a sledge acts partly down along the slope, and partly downwards at right angles to the slope; the force acting downwards at right angles is met by an equal but opposite reaction from the slope itself, but the force acting down along the slope drives the sledge and overcomes the friction between the runners and the slope. Call the force acting down the slope the *driving* force, and the force acting at right angles to the slope, the *sinking* force. How much bigger or smaller one of these two forces is than the other depends, of course, on the angle of the slope.

We can apply this picture to a gliding pheasant if we imagine each of its wings to rest on a smooth rigid runner sloping downwards towards the earth (Fig. 46); as with the sledge, the weight of the bird would give a 'driving' component and a 'sinking' component. As the bird began to slide down the runners, the motion of its stiff wings through the air would induce lifting forces against the wings, and because of this the passive reaction from the rigid runners would decrease; at the same time, the wings and body would be subjected to a 'drag' force acting along the line of the runners in a direction opposite to that of the 'driving' force of gravity. As the speed of glide increased, the lift would also increase, until a time would come when the lift is equal to the 'sinking' force of gravity, and the drag force is equal to the 'driving' component. At this moment we might take

124

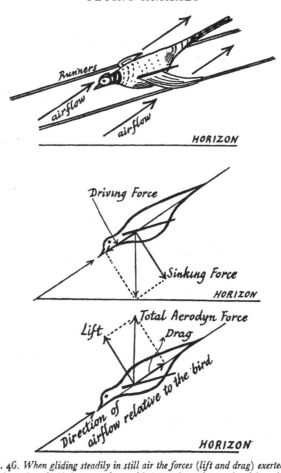

FIG. 46. *When gliding steadily in still air the forces (lift and drag) exerted by the air against the bird are exactly equal but opposite to the weight of the animal. The conditions are comparable with those which would exist if a bird were sliding down a pair of rigid runners in a vacuum: the reaction from the runners at right angles to their surface would be equivalent to the lift of the wing and the friction of the runners to the drag of the bird's body*

away the rigid runners without making any difference at all to the bird: it would go on gliding through the air at constant speed.

A good glider travels a long way horizontally with the smallest possible loss of height, and you can judge how good it is as a glider by measuring the angle between the track of its motion and the level horizon. It is very important to note that this angle does *not* depend on the weight of the bird; it depends solely on the ratio of the forces – the 'lift' force to the 'drag' force – exerted by the air; that is, it depends on the shape of the wings and on the angle which the surface of the wings makes with its own direction of motion. The speed of the glide, on the other hand, depends partly on the weight of the bird and partly on the size of the wings. A heavy bird with small wings glides rapidly; a light bird with large wings glides slowly.

The distance that can be travelled by a bird when it is gliding through still air under its own weight is, of course, limited by the height from which the bird starts – sooner or later it must reach earth and the glide must come to an end. But if the air is not still, but is itself moving over the earth, the story is very different. The lift and drag forces exerted by the wings depend solely on the particular motion of the wings through air; if the air itself is moving, the motion of the bird relative to the earth is the resultant of these two motions: the motion of the bird through the air, and the motion of the air itself over the ground. For example, if a bird is gliding down an air slope losing vertical height at a rate of 5 ft per sec., and the air itself at the same time is rising from the ground at 5 ft per sec., then the bird will glide along level above the ground, although it is all the time travelling downwards through the rising air (Fig. 47). And if the air is rising from the ground faster than the bird is losing height through the air, the bird will keep on getting higher above the ground although it makes no wing movement to do so. A horizontal wind cannot affect the rate at which a bird loses height above the earth – such a wind affects only the rate at which the bird moves along horizontally over the ground during its fall. What will happen then, when the wind is blowing backwards and upwards at exactly the same speeds as a bird is moving forwards and downwards along its cushion of air? Seen from the ground then, the bird remains 'fixed' in space (Fig. 48). In such conditions a bird is not unlike a man who is walking *down* a moving stairway at the same speed as the stairway itself is moving upwards.

The extent to which upward currents of air may account for the behaviour of gliding birds can be judged by watching birds at flight in regions where upward air currents are known to exist. There are two main causes of upward air currents: first, when a horizontal wind meets an obstruction and is deflected upwards; and secondly, where air, warmed by the surface of the earth, moves upwards and is replaced by colder air from high levels of the atmosphere. Typical cases of gliding on up-currents caused by obstructions may be seen when swifts glide along the eaves on the windward side of a building; or when gulls glide along a line of cliffs when the wind is blowing on-shore; or when eagles and buzzards glide on the windward side of mountains. When a strong head-

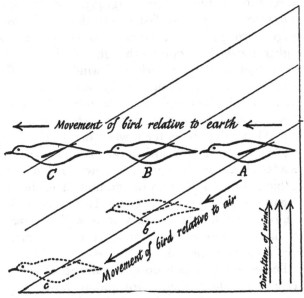

FIG. 47. *When gliding in still air a bird is constantly losing height relative to the ground, but if the air is rising from the ground as fast as the bird is falling, it is able to glide along a horizontal path for an indefinite period*

wind is deflected upwards by a mountainside, buzzards may hang on the air, fixed in space high above the earth for surprising lengths of time.

In temperate climates the use of thermal up-currents by gliding birds is somewhat restricted; nevertheless, birds may sometimes gain considerable height over towns or other regions where warm air is rising. In

tropical countries, however, cheels, vultures, and birds
of this sort, are very good at this kind of glide (Fig. 49).
During the mornings, in these places, columns of
warm air rise from the ground and drift downwind.
The birds glide down a spiral within such a column of
air, and as the rate of their downward movement is

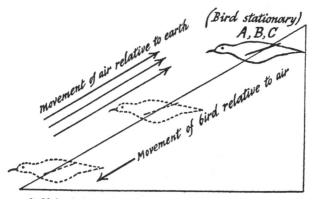

FIG. 48. *If the air is moving backwards and upwards from the ground at the
same speed as the bird is travelling forwards and downwards, the bird can hover
with motionless wings, quite stationary to an observer on the ground*

less than the rate at which the column itself is moving
upwards, the birds constantly spiral upwards, getting
higher above the earth as they drift downwind. As we
might expect, thermal gliding is characteristic of birds
with light bodies and large wings. On these two factors
(weight and wing surface), depends the speed at
which the bird would lose height when gliding in still
air and, consequently, the speed at which the air must
rise if the bird maintains a horizontal or upward path.

So far, the principles we have applied to gliding

birds are exactly the same as those made use of by a glider pilot, whose powers of gliding and of staying in the air depend almost entirely upon the use of obstructional and thermal up-currents. Certain birds, however, are able to glide in conditions where there seem to be no up-currents: shearwaters and albatrosses, for

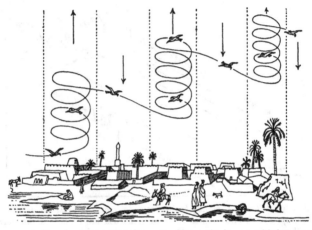

FIG. 49. *In regions where upward currents of warm air alternate with downward currents of cool air, birds can glide upwards in the former and glide downwards in the latter. Arrows show direction of air movement. It is seen that a bird can gain considerable height by taking advantage of these air currents*

instance, glide near the surface of the sea, though how they do so no one has yet managed to explain. Physical research may one day show us that up-currents of air arise near the leading surfaces of travelling waves, and that the surface gliders are able to use these currents. Or it may be that these birds have solved the problem of 'gust' flying – gaining height when they meet a gust, and losing height again as the gust dies away, rather as a tennis ball can be kept in the air by a succession of small upward hits from a racquet.

All observers agree that an albatross cannot maintain its gliding flight if there is no wind. These great birds seem to rise in a slanting direction against the wind, to a height of about 20 ft, depending on the strength and direction of the wind; this climb is followed by a wide semicircular turn as the albatross rapidly descends downwind; and then the cycle is re-

FIG. 50. *The flight of the albatross. Note how the bird makes semicircular turns, climbing into wind and descending downwind*

peated (Fig. 50). As the bird turns into the wind again, its speed relative to the air has been increased and the kinetic energy (energy due to movement) thus gained can be used for gaining height; this results in a gain in potential energy (energy due to height) which is then available for maintaining the bird's flight downwind; and so it goes on. Notice that this interpretation of the flight of the albatross depends on the existence of an adequate difference between the wind speed at the surface of the sea, and that at a level perhaps 20 ft up.

Gulls behind a steamer give us an example of gliding that is by no means as simple as it may appear. No doubt, there is an upward deflection of air by the hull of a ship, and the heating effect of the vessel's smokestacks may cause certain upward thermal currents, but there is also a mass of air, stationary in relation to the ship, and therefore travelling over the sea at the same speed as the ship. How far the gulls rely on the up-currents, and how far they glide down in the 'stationary' air, would form an interesting subject of study.

Stability in flight is of the greatest importance in an aeroplane which must be designed so that any slight disturbance in balance between lift, drag, and weight is automatically and immediately corrected, yet all the conditions of balance must also be under the command of the pilot, for by control of these he must fly, and fly as he decides to. Similar problems must be solved by birds. A bird achieves control partly by using the tail as a horizontal rudder, and partly by changing the shape and posture of the wings; and these movements are also under the immediate control of the semicircular canals of the bird's ears, just as are the movements of the fins of a fish.

6

Flapping Flight

A BIRD flying – propelling itself by its own muscular efforts – is one of Nature's great masterpieces, and she guards very closely the secrets of her success. It is easy to see that the bird's wings are beating upwards and downwards and to realize that these movements must provide a lifting force equal to the weight of the body, and a forward thrust equal to the backward drag of the air. From a mechanical point of view, the wings of the flying bird are carrying out simultaneously the functions performed respectively by the wings and the propeller blades of an aeroplane; a bird's wings are, therefore, more closely comparable with the rotor and screw blades of an autogyro or helicopter.

But any attempt to make an accurate study of the movements of a flying bird meets with difficulties. The form of the wing, for instance, is constantly altering during the course of its beat, some of the changes being due to the suppleness of the wing feathers and others to the bird's own internal muscular movement. Then again, the bird's speed and the frequency of its wing beats are too great for the wing to be exactly observed by the human eye. As a general rule, the smaller the bird the faster and more often it beats its wings; a swan or a heron beats its wings about two times, a gull five times, a pigeon ten times, and a humming bird fifty times per sec. Details of such wing movements can only be found out by high-speed cinematography using a special light that gives a succession of very bright and very quick flashes. The photographs shown on Plates 11 and 12 were taken by Dr R. H. J. Brown, in this way; they show the movement of the wings of a pigeon flying slowly across a room.

In Plate 11 a pigeon is flying towards the camera. The down stroke begins as shown by photograph 1 with the wings fully extended, stretched up over the back of the bird. Both wings then strike downwards (photographs 2–4) with the whole surface of the wing moving down almost at right angles to its horizontal path of motion; the primary feathers are bending upwards under the pressure of the air (photograph 3). When the wings have reached a horizontal position (photograph 4), the downward motions cease and the wings swing forwards, to meet in front of the body (photographs 6 and 7); during this change of movement the primary feathers separate from each other, and bend sharply upwards under the lifting effect of the air moving between them. Photograph 7 shows the end of the downward and forward stroke of the beat. The backward and upward stroke starts with an upward and backward movement of the wrist joint, uncovering the head and leaving the primary feathers directed forwards along the line of flight (photograph 8). Then the primary feathers are drawn violently backwards and upwards (photographs 9–11) and the whole wing straightens out in position for a new downward movement.

These same phases of a complete wing beat can also be seen from another direction in Plate 12, where the pigeon is flying across the field of the camera; photographs 1–6 show the downward and forward stroke, and photographs 7–12 show the upward and backward stroke. Photographs 2–4 show the separation and bending of the primary feathers during the first moments of the down stroke, and the forward swing of the whole wing is seen in photograph 6. The raising of the elbow and the subsequent very rapid backward flick of the wrist and primary feathers are seen in photographs 8–11. From these photographs it is possible to draw three conclusions with fair certainty: (i) During the initial downward wing movement the wing surfaces are travelling down in a direction almost perpendicular to their path of motion, and consequently elicit a powerful upward reaction from the air. (ii) During the second part of the down stroke the wing is travelling forwards with its surface inclined to the path of motion at a relatively small angle; it thus acts as an aerofoil and develops a lift in the normal way. (iii) The main propulsive thrust comes from the rapid backward movement of the primary feathers, which occurs during the upward stroke. If this analysis is correct, it may be said that the down stroke of the wing gives the lift and the up stroke produces both lift and forward thrust.

But can a bird's wing do so much, and do it constantly, without getting tired? The down stroke is easy, for the wing is pulled downwards by the very large *pectoralis major* muscle; but the muscles that move the wrist and raise the wing are small by comparison, and it is difficult to see how a very rapid series of up strokes could be kept up. And, in fact, it cannot. The photographs shown in Plates 11 and 12 were taken during slow short flights of about 20 ft, and it soon appeared that pigeons find such flights tiring; eight or nine flights left this bird disinclined to take further part in the experiment. It seems likely, therefore, that the wing movements observed during and immediately after a take-off from the ground are not the same as those of a pigeon in full sustained flight in the open air, for we know that such birds can in fact travel for a long time without showing fatigue. Dr Brown, therefore, decided to photograph the bird in full flight down a long corridor. A series of such photographs is seen in Plate 13. Two main changes are to be noticed in the down stroke: (i) the amplitude of the beat is smaller – the wings no longer rise high enough to meet over the bird's back; and (ii) they no longer swing forwards at the end of their downward movement: the bird's head remains clearly in sight at all stages of the flight. Changes also occur in the up stroke; the backward flick of the wrist and primary feathers has almost, if not completely, disappeared: the inner surface (radio-ulnar) of the wing begins to rise first and is followed by the wrist and primaries. What these characteristic features of open flight may signify is not altogether clear. During the down stroke, the whole wing travels downwards and forwards at a relatively small angle to its path of motion, and it seems likely that the resultant 'lift' is directed both forwards and upwards, and can therefore propel the bird as well as support its weight. On the other hand, the up stroke of a bird in full flight appears to be of a more passive kind than the up stroke immediately after a take-off; perhaps during free flight the wing is passively raised by the pressure of air against its under surface, and its rate of upward movement controlled by the braking effect of the *pectoralis major* muscle. If this is so, the wings, during the upward stroke, act as kites to lift the body merely at the expense of some diminution of forward speed – rather like an umbrella opened downwind on a windy day. Whether this suggestion will bear further scrutiny remains to be seen, but at least it seems to explain one way by which rapid free flight can be made much less tiring than a take-off from the ground – for if our explanation is correct, the relatively small *pectoralis minor* muscle would be in action only for a little while after each take-off. After that the only muscle concerned would be the large *pectoralis major*.

Among vertebrate animals there is only one other – the bat (Plate 16 and Fig. 51) – whose powers of flight come anywhere near those of a bird. At present, we know little about the movements of a bat's wings, but as far as we can see, they appear to be similar to the motion of a bird's wings, and it seems likely that the same general principles would apply to both. What is characteristic of a bat's flight is, however, its ability to avoid obstacles when flying in the dark (Plate 17). At one time, this very useful faculty was believed to be due to sense organs on the wings capable of registering the slight alterations in air-pressure which must follow when the moving wing came too close to a solid obstacle. It has been known, however, for many years that a deaf bat frequently collides with obstacles. It has now been discovered that the vocal chords of a bat can give out sound-waves of very high frequency, far

FIG. 51. *A bat in flight. The wing action is similar to that of a bird*

higher than those to which a human ear can respond: these waves – not to be confused with the 'squeak' of bats which some people can hear – are of 30,000–50,000 vibrations per sec., and there can be little doubt that a flying bat sends out these 'supersonic' waves and listens to their reflection from surrounding objects. In effect, the bats have evolved a very efficien echo-sounding equipment. We can make a model of such a system. A 'Galton whistle' capable of sending out sound-waves at 50,000 cycles per sec., is mounted underneath (but shielded from) a telephone capable of responding to waves of this frequency. Then, if no obstacle lies in the path of the whistle's outgoing high waves, the telephone will not respond. But if any object is present that can reflect sound, the outgoing waves will be reflected back and picked up by the telephone where they can be made audible to an observer by means of an instrument reducing their frequency to a level to which human ears will respond. A model of this kind will instantly detect the presence of a glass window.

Among the invertebrates, only insects appear to have solved the problem of sustaining themselves in the air by means of wings. In most cases, a lift equal to the weight of the insect's body only develops when the wings are forced through the air by the insect's own muscular effort; the red admiral butterfly is one of the few insects capable of gliding flight. As with birds, the smaller the insect the more frequently does it beat its wings. A swallowtail butterfly beats about five times, a hive bee or a horse fly about 200 times per sec. The record, so far as is known is held by the males of a small midge (*Fercepomyia*) whose wings vibrate more than 1000 times per sec. and produce a very high-pitched note. Muscular activity of this order requires a great deal of energy, and most insects can only maintain active flight if they are reasonably warm and supplied with plenty of food. We can measure the 'fuel' (or food) requirements by allowing a small fruit fly (*Drosophila*) to beat its wings until all the food reserves in its body are exhausted: this will be after 2–4 hr of flight, according to the age of the fly. Flies wearied out in this way, can resume flight about 30 sec. after being given a drink of sugar solution. Sustained flight is possible as long as the fly is provided with 3–10 mg. of sugar per min.

For an insect to fly along a level path, its wings must move in such a way as to exert a downward and backward thrust against the air; as the insect moves levelly forwards, the air is driven downwards and backwards. We can see the air moving in this way in Plate 18. The fly is photographed from one side against a dark background. The wings are vibrating, and catch the light at the top and bottom of their beat, and while the wings are beating, a shower of very fine lycopodium particles is allowed to fall on the fly from above and a photograph of $\frac{1}{25}$ sec. exposure is taken. The path travelled by a particle in this short time is shown as a white streak in the photograph, and the length of the streak gives the speed at which the particle is moving. You will see in the photographs that they enter from above the zone of the wings at a relatively slow rate, and are forced backwards and downwards out of this zone at a much higher speed (in this particular case about 6 ft per sec.). An insect's wings, as a free-flying bird's wings, do the work at one and the same time, of propeller and wings of an aeroplane.

An ordinary house-fly weighs about 14 mg. or about $\frac{1}{2000}$ oz., and a force equal but opposite to this weight represents the 'lift' derived from the wings during level flight, while the backward component exerted by the wings against the air is equal but opposite to the 'drag' encountered by the fly as it moves forwards through the air. Some of the ways in which these forces are kept in balance with the weight of the body are being investigated by Dr Hollick at Cambridge by means of a very ingenious and very delicate balance. This consists essentially of a very thin glass rod fixed firmly at one end so that it extends horizontally. The fly is mounted at the unfixed end of this rod, and any horizontal backward thrust developed by the fly causes the rod to bend to one side, and any 'lift' force causes the rod to bend up. The magnitude of the forces can be determined by measuring the amount to which the rod bends sideways and upwards. Any movement of the rod is detected by the reflection of light from two small mirrors attached to the rod, which reflect light on to a graduated scale.

As with a bird, the flight of a fly must be stable in the sense that accidental disturbances in the balance of lift, drag, and weight must be quickly or automatically corrected. Precisely how this is done is not known, but it is quite clear that insects possess special sense organs which inform the brain of changes in the direction of motion of its body. Grasshoppers have special hairs on their heads, and it is the 'feel' of the air as it flows past these hairs that enables the insect to keep its head facing into the direction of its flight. Dr J. W. S. Pringle has investigated a particularly interesting organ of direction control which exists in the two-winged insects (*Diptera*) or flies. In these insects the two hinder wings are replaced by two small knobs or 'halteres' united to the body by stalks (Fig. 52).

During flight the tips of the halteres swing to and fro in the arc of a circle. When the fly is turned off its course, the halteres continue to swing in the same

FIG. 52. *The hind wings of flies are modified into club-shaped structures known as halteres (indicated by arrows). The halteres vibrate up and down and act as two-dimensional gyroscopes, enabling the animal to fly along a steady path*

plane as before the turn, on the principle of a gyroscope, and the base of the haltere stalk is consequently twisted. This twisting excites a nerve, and the brain thereupon sends appropriate instructions to the muscles that control the wings.

By this time you will know something about how birds take to the air and how birds, bats, and insects maintain their flight. When the flight ends, how do they return to earth? Aeroplanes, as you know, must be able to land as well as to fly, and so must flying animals. Legs are the landing carriages of flying animals; as the flier comes in to its landing place its legs are brought into position ready to take over the weight of the body. At the same time it is necessary for the forward motion of the flying body to be brought to an end at the moment of landing. Watch large birds alighting and you will see clearly the whole process. In free flight they carry their legs out of the way – usually stretched out behind them; but as a bird comes in to land the legs are lowered so as to be there to take the weight at contact. At the same time, the wings are tilted backwards, thus reducing the lift and greatly increasing the drag force exerted by the air. Almost as soon as the bird's feet touch ground, the wings furl neatly at the sides of the body. The movements of wings and feet are beautifully co-ordinated – not only at landing but also at taking-off – for the spread and beat of the wings is associated in some way with any reduction of pressure between the feet and the foothold. This can be easily demonstrated by suddenly lowering the perch on which a bird, such as an owl, is standing: the wings immediately expand and beat downwards.

The legs and wings of insects are related in an equally striking way. If the back of a fly is fixed by a small drop of wax to a fine glass rod, the wings will remain at rest so long as the legs of the fly are allowed to rest on some firm object. If this object is withdrawn the wings often begin to beat at once (and will always do so if a current of air is blown against the fly's body), and when the wings beat the hindlegs are drawn up and back and the front legs are held out in front, so as to leave free a zone for the beating wings.

And now we come to the end of our story. I have tried to tell you how some of our commoner animals move and how they obey the same fundamental laws as those which govern the movements of inanimate things. The picture I have given is necessarily brief and incomplete, but it may, I hope, help you to look at moving animals with greater interest and understanding than in the past, and perhaps encourage you to make observations of your own. The more we learn about living animals, the more beautiful their activities appear; and in a human world so much concerned with ugly engines of destruction, the concept of natural beauty provides welcome relief.

Printed in the United States
By Bookmasters